Hugo Dixon

About the Author

WILL STORR is a feature writer and travel journalist based in south London. He has been voted New Journalist of the Year and Feature Writer of the Year, and his work has appeared in a wide range of titles, including *The Times*, *The Observer*, *The Mirror*, *The Face*, *Arena*, *High Life*, *Q* and *Loaded*. To read more of Will's work or to contact him, visit www.willstorr.com.

WILL STORR

VS.

THE SUPERNATURAL

WILL STORR

VS.

THE SUPERNATURAL

ONE MAN'S SEARCH FOR THE TRUTH ABOUT GHOSTS

HARPER

NEW YORK • LONDON • TORONTO • SYDNEY

HARPER

The prologue originally appeared as a feature in *Loaded*.
I am grateful to Andrew Sumner for granting me permission to reprint it.

Chapter One originally appeared as a feature in
The Times Magazine. I am grateful to Simon Hills
for granting me permission to reprint it.

Excerpt from Canon Michael Perry's *Deliverance*,
published by SPCK, appears by kind permission.

Excerpt from *This House Is Haunted* appears by
kind permission of its author, Guy Lyon Playfair, and Souvenir Press.

Quotes from the *British Journal of Psychiatry* appear by kind permission.

Quotes from *Journal of the Society for Psychical Research*
appear by kind permission.

The website that appears on page 84 is copyright Fujifilm.
Thanks to Jenny Hodge for granting me permission to reprint it here.

The names of some incidental characters have been changed.
Some small segments have been relocated.

Originally published in the United Kingdom in 2006 by Ebury Press,
an imprint of Random House.

FIRST HARPER PAPERBACK PUBLISHED 2006.

Designed by seagulls

Library of Congress Cataloging-in-Publication Data is available upon request.

ISBN-10: 0-06-113219-5
ISBN-13: 978-0-06-113219-3

06 07 08 09 10 ❖/RRD 10 9 8 7 6 5 4 3 2 1

For Farrah

Contents

Prologue	'It puts a scar on your brain'	1
1	'Are you Sir Thomas Sackville?'	22
2	'It's not all coming from the trees'	39
3	'Strange patterns'	48
4	'Come back, Rain-On-Face'	64
5	'Distrust the mystic'	75
6	'Making things fit'	85
7	'All I ask is that you put your life in my hands'	102
8	'Turn the light off, bitch'	116
9	'I was very upset at what I saw'	124
10	'Open your eyes'	137
11	'I promise you, you'll scare yourself'	148
12	'They'll build it up and bugger off home'	162
13	'We've got strangers in the house'	180
14	'They called me Ghost Girl'	186
15	'That's Annie's room'	200
16	'And when they die, they'll get a big surprise'	213
17	'Some really weird things'	226
18	'Kangaroo!'	239
19	'I talk to the devil every day'	250
20	'I am the ghost'	263
21	'And that's God?'	284
Epilogue		310
10 Most Haunted Places in America		312
Acknowledgements		322

WILL STORR

VS.

THE SUPERNATURAL

Prologue

'It puts a scar on your brain'

Don't freak out. That's rule number one. I take a second to absorb this information, before writing it carefully on page one of my reporter's notebook.

'You got that?' says the demonologist. He peers at me over the rim of his coffee cup. 'It's extremely important.'

'Rule number one, don't freak out, yes.' I nod and look down, as if to make sure that I did, in fact, get it. 'Got it.'

'Good. When I start asking questions, there may be things going on around you. Whatever you do, *do not freak out.* If you feel that you have to get out, then just walk right out of the house.'

He pauses and takes a long drag from his freshly lit Kool Mild. The smoke drifts out of his mouth and floats up towards the slow, dirty ceiling fan.

'Just say, "I've got to go for a cigarette", or "I gotta get some air", something like that. OK?'

'Gotcha.'

The demonologist takes another sip of his coffee. He looks at me blackly from under his eyebrows.

'Rule number two. When we start the investigation and we turn the lights down, try not to move.'

'Don't move.'

'Right. Do not move or adjust yourself unless it is absolutely, positively necessary. We're looking for something in the house that's paranormal ... '

'And you don't want to get the sound of paranormal stuff confused with the sound of me adjusting myself?'

'Right,' he says.

'Um – another coffee?'

'Sure, great,' he says, and I switch on my best waitress-catching face. 'We're gonna need it.'

I'm in a roadside diner somewhere on the outskirts of Philadelphia, USA. Everything that just jumped into your head when you read the words 'roadside diner' and 'USA' is actually here. There's a run-to-fat waitress with a heart of gold and a hairy chin, chewing gum and taking orders for cawfee. There are little, table-top jukeboxes with slots for quarters and multiple Elvis options. There's a handsome stubbly man, who looks like he's on the run from the law, sitting at the counter, chewing a toothpick and considering his next move. Any minute now, he's probably going to start a fight with the serial killer in the next booth.

I'm here on a journalistic assignment for *Loaded* magazine. Two weeks ago I was at my parents' house, perusing my brother's *Web User*. And as I flicked, I noticed that the online home of a real-life ghostbuster had been awarded 'Website of the Month'. His site looked fantastic. Encyclopaedic, grandiose and full of gothic kitsch and portentous admonitions about divination and devil worship. This, I thought, would make a brilliant story. It'd be fantastic, because it concerns an American eccentric, and American eccentrics are great. They're more sincere, unabashed and convinced in their madness than any other eccentrics in the world. And they say hilarious things like, 'Rule number one is don't freak out.'

Lou Gentile is thirty-two, married and a proud father of two. He's eager, open and upbeat, with the Italian looks and confident countenance of a Mafioso gone to seed. By day, he is a central heating engineer. And by night he is the demonologist. An investigator into, and crusader against, the evil forces of the demonic. Amongst other treats, Lou has promised to show me 'ghost lights' on his infra-red monitor and play me 'electronic voice phenomena' – EVP – on his Dictaphone. These bizarre growling, grunting sounds are, apparently, the noises of spirits trying to communicate from another dimension. When I called on him an hour ago, he was watching *Ghostbusters II* with his young daughter on his billboard-sized TV. And tonight, he's driving me to a haunted house that was built on top of a graveyard. So far, as you can see … everything is going exactly to plan.

'Rule number three. Let me know what you see, what you feel and what you hear. If you're sitting there and you feel extreme coldness coming over you, then something's going your way. You need to tell me. Some of this stuff can be explained easily, draughts and stuff, but you'll know. It's hard to describe, but you'll know when it happens. And rule number four,' he says, tipping down the dreggy drips of another cup of coffee, 'is don't laugh. It is very important when you're going through something like this that you don't laugh. And that can be hard.'

'Right,' I say. I glance to my left through the window. There have been blizzards all down the eastern seaboard for the last three days. Fat whacks of snow cover the ground everywhere except the freeway. I pause for a second to watch the cars and trucks and monstrous articulated lorries bomb noisily through the night, all exhaust-steam and slipstream and white lights and red. And as I sit and look at the traffic, somewhere deep in my brain, a tiny alarm starts to sound. At this moment, I'm still barely aware of it. But I've just begun to sense that *something* isn't right.

We pay our waitress and walk towards the door. As I zip my coat in preparation for the freezing air, I ask Lou what'll happen if I don't follow his rules.

'You could get kicked,' he says distractedly, as we pace through the car park, 'or hit by something flying across the room. Anything could happen. It can get extremely bizarre.'

'Oh,' I say. And the tiny alarm in my brain gets just a little more insistent.

'Lou,' I say when we reach his Lincoln Continental. I open

the door and climb into the cold, leather passenger seat. 'I'm not in any danger, am I?'

Lou chugs the ignition and turns the heating dial to full. The headlights flash up and hit the wall of the diner in front of us. Large flakes of snow amble in through the beam, from the darkness all around it.

'You're a Roman Catholic, right?'

'Yep.'

'If anything happens, if you get freaked out, just imagine yourself in what's called a white Christ light. There's nothing that can penetrate that light. That's your faith.'

'OK,' I say. My eyes flicker down. 'Right.'

A problem. I lost my faith in the middle of my R.E. A-level. That's if I had any in the first place. In theory, with the upbringing I was put through, I should be bloated with glowing, golden faith. My parents, you see, were full-time, professional Catholics. My mum was chairman of the governors at the Catholic comprehensive that I attended and, on top of that, she'd done some sort of course that enabled her to give out communion at Mass. My dad, in his leisure time, was in charge of the choir and organ-playing at church, and used to go on fun Gregorian Chant weekends at a monastery in Belgium. His entire working life was spent in the service of Catholic education. To reward his five-star devotion to the cause, an archbishop presented him with a certificate at a Mass on his retirement. It was signed by the Pope himself. Really. The Pope.

To me, though, religion never quite added up. I used to be taken to church against my will every Sunday to listen to the terrifying priests with their incense, magic and bad news. And I could never work out how that did me any good *whatsoever*. More importantly, I couldn't see what good it did God. He must be quite an insecure fellow, I'd think, if he needs all this reassurance that we like him. I couldn't understand how He could be merciful and loving on the one hand then send me to hell for doubting his existence the next. And I'd look around the church every

Sunday, at all the pursed lips and the hoisted chins, and I'd think, *These* are God's people? *These* are the ones who've got it right?

As I got older, I started to learn more about the contradictions, the baffling rules and the surreal illogicality of the Church. And as I became more questioning of the constant low-level doom, the funny costumes, the weird tales and the rank hypocrisy, increasingly, I'd just think … *naaaaah*.

Eventually, I became a proud hard rationalist. I can remember the moment it finally happened, during an A-level R.E. lesson at school. I'd just been taught that the Bible hadn't, as I'd been led to believe for the past eighteen years, been written by God after all. No. The old bit, we were told, was a collection of stories some ancient nomads used to tell. And the new bit was authored by a group of angry political activists with scores to settle. Everything I'd been taught was a lie. In that instant, I turned my back on every flavour of the supernatural. Religion, the afterlife, ghosts, spirits, the lot. To me, it was all so much superstition. It was absolutely clear – people are desperate to believe in whatever will comfort them through this chaotic, random and, ultimately, pointless life. And if someone wants to convince themselves that there's an afterlife, either by talking to God or believing in ghosts, then they're welcome. As far as I was concerned, though, existence was just a happy accident. Up was up, down was down, and when we died we were nothing but a bag of old guts.

So, when I told Lou that I was a Roman Catholic, just like him, it wasn't a 100 per cent lie: legally, on paper, I *am* a Roman Catholic. But if the only thing protecting me from having an evil demon throw a chair at my head is my 'faith' …

'You OK, Will?' Lou asks, clearly noticing the glaze that's set over my eyes. 'You worried?'

'No,' I say. I manage to snap myself out it and look ahead into the freeway. 'No, just wondering what … what … where are we going?'

'Bishopville, Maryland. These people are suing a property developer because the land he sold them, which they built their house on, used to be a graveyard. And they don't want to live on top of a graveyard.'

'I don't blame them.'

'It's pretty bad, yeah. As you look at the house, you can actually see the pits. All the headstones were found in the river.'

'And they're having problems with ... ?' I prompt.

Lou reaches down and pushes the car's cigarette burner in. 'The TV turns on and off by itself. Lights go on and off. Doors slam shut at night. The wife says that she has seen these horrific faces in the glass. The son's had his room dismantled, things have been thrown about, he's seen a black shadow, an image of an old man in his room, different things like that.'

I watch as the demonologist picks a Kool Mild box off the dashboard and lights another cigarette, his gaze never flitting from the snowy, rushing road in front of him.

'It's definitely haunted,' he says. 'I don't know how much activity we're going to experience tonight, but I know for a fact it's haunted.'

'What have you experienced there?' I ask.

'Noises. I got a couple of very odd pictures. Now, tonight, for the first time, we're going to use the audio recorder and see what we get out of that.' He takes a long drag on his cigarette and indicates to pull out in front of a red pick-up. 'You know, I can't guarantee that you'll see anything tonight. It just don't happen like that. You can go five cases without experiencing anything and then, on the next one, see all the shit in the world. And, believe me, you don't want to see some of the shit I've seen. It puts a scar on your brain.'

We drive on, with our thoughts, as the miles rush behind. Illuminated road signs loom and reach over us. Trees, towns and turn-offs buzz past, all anonymous and alike, obscured by the dark and snow. After a time, Lou begins to talk. As he does, orange and pale light from the road runs over his face and his eyes dart quickly from tarmac to bumper to road sign and back. He tells me that his mission, as a demonologist, is to gather evidence on behalf of people who are suffering from a perceived haunting. He then compiles a dossier, which he'll present to a priest in order to persuade them to provide the hauntee with an exorcism. As the Church becomes more modern and PR-conscious, its clergy are increasingly reluctant to perform this spooky, retrogressive and secretive rite. They demand cast-iron evidence before they'll commit. And, since suffering a paranormally troubled childhood, Lou feels he has a 'calling' to gather this evidence on behalf of the terrified, helpless and disbelieved.

'I had a recent case, up in this attic,' he tells me. 'This was a very, very stale place. A very oppressed place. I went up there and I could hear all this stuff moving around me. And then the smell came. A stench. Standing right in front of me, I saw one of those black shadows, about eight or nine foot high. My back was against the window and it starts breathing heavy and I'm looking at it like, that's not my breathing, and then – and I don't know why I said it – but I said, "You know what? You really sound like you should be dead." With that I got pushed up against the window and I almost went through it. Well, I had to tail-ass it out of there. You know, I'm not Superman.

'I was there when the priest was exorcising that house. He did three, the same night. The first and the second ones went OK. The third one, he started fumbling around with his words. The client started freaking out. Things were starting to vibrate and almost move, just like in the movies. Then the priest, I watched his cross levitate and almost get pulled off. Then he stopped and composed himself and he started to say it again. Everything was real quiet until he said the final word. That's when this loud growl came out of nowhere.'

'Oh dear,' I say.

'I damn near shit my pants. It sounded like a ... I can't even describe it. Like a lion almost, but it's something that's not human or animal. That was a bad one. There were screams and there were three knocks during that. Always three.'

'Why three?' I ask, as Lou palms the steering wheel right and we pull off the freeway.

'Threes are a blasphemy to the trinity,' he says. 'Father, son and holy spirit. When these things come in threes, it's mocking God. Often, you'll experience stuff between three and four in the morning – 3.33 a.m. can be a real hot time. Hey Will,' he says. A new thought has distracted him mid-flow and he turns, quickly, to meet my eye. 'You live in London, right?'

'Yeah.'

'You must have heard of the Enfield poltergeist, then. Right?'

'No. What's that?'

'Oh, man,' says Lou. He shakes his head, rolls his eyes and lets a small smile curl one side of his mouth. 'The Enfield case was just insane. One of the biggest, best-documented poltergeist cases in history. A real bad demonic case. Man, you should check that one out.'

We pull up to the driveway of the big, white clapboard house. Two full-sized dogs bound up to shout at the car. I look around. There's a makeshift skateboard ramp and a basketball hoop and a green hose coiled up on the grass. As he opens his door, Lou tells me that if I want a cigarette, I have to go outside as the Carvens have a no-smoking policy in the house.

And they really are no-smoking-in-the-house kind of people. Tom and Deborah Carven are a bright, smart and attractive couple in their late thirties. They welcome us in, warmly, as their obediently brought-to-heel dogs leap and lap up around their waists. The worry and embarrassment that they're obviously feeling at the arrival of a professional demonologist and a British journalist into their otherwise picket-fence perfect lives is almost fully suppressed by their impeccable, smiley manners. They have two cars, two teenage sons, designer stools by their breakfast bar, 'girl scout cookies' arranged neatly on a plate and between twenty and thirty bodies buried in their front garden, according to legal depositions from local farmers who once worked the land.

'Our youngest son, Timmy,' says Deborah, as we settle down in the large, cosy lounge, 'from the time he could verbalise, used to talk about this old man that would show up and talk to him.'

'We just thought he had a wild imagination,' Tom says, sitting forwards on the armchair next to her, 'but he kept seeing it and he'd have all these different stories and stuff. He was three or four at the time, afraid to sleep with the light off, talking about "the old man down the river".'

Deborah pulls her feet underneath her legs on the sofa. She's pretty and immaculately, Americanly groomed, with bobbed brunette hair and a baggy, white cable-knit jumper.

'And then we had the TV switching on and off, lights blowing,' she says. 'The back door would slam two or three times during the night.'

'We'd both sit up in bed right at the same time,' Tom says. 'Of course, I'd grab the old gun and run downstairs ... nothing.' He shrugs.

'My worst memory is this one night I couldn't sleep,' Deborah says, with both hands wrapped around her mug. 'I came down here and I laid on this sofa and I made the German Shepherd come and lay down beside me. I just had this weird feeling. I was lying there and on the exterior of the glass was these, like, skeleton ghoulish heads and I'm just like ... '

Deborah Carven shakes her head at the memory. Her husband stares, with his elbows resting on his legs, towards the rug. Lou potters in the background setting up tripods and laptops and cameras. These people *must* be crazy, I think. They must be.

'So, what do you do for a living?' I ask Tom.

'I work in a prison in Georgetown, Delaware. I'm a recreation specialist.'

'And I own the local deli,' says Deborah. 'That's where I first heard about the graveyard. All the farmers and locals come in to sit and have coffee. One morning, one of them started telling me that we live on a graveyard and I'm like, "Excuse me?" So, one night I got a shovel, and I felt a real idiot. I thought, I'm like, nuts. But I started digging and I pulled out a bone. I was like, OK, it could be a dog, it could be a cow. Then I started digging with my hands. I pulled out a femur and a hip joint and a man's medallion. And then I pulled a handle out, from a casket.'

There's a silence. Deborah looks at me with sad, uneasy eyes.

'I just went, "That's no cow."'

By now, it's after midnight and Lou's almost ready to begin. All the lights are off. An infra-red camera is pointing up the stairs, towards the landing outside Timmy's room where the old man appears. And the monitor that's attached to it is sitting on a coffee table, along with a Dictaphone and a digital thermometer. I check the time on the clock on the wall. It's stopped on the stroke of three o'clock.

And then, just as I expected, nothing happens. For hours. We sit in the dark and watch the monitor, which has bathed the room in a weird spectral glow. The only sounds in the house are the odd structural creak and the shuffling, sniffing and coughing sound of humans trying to keep still.

After a couple of church-length hours, Deborah whispers that she's cold and is going upstairs to fetch a blanket. We watch the green screen as she walks up, hand on the rail. And then we see it, all three of us. A luminescent white globule, about the size of a fist, appears floating above her head. It follows her up the stairs and then moves, slowly and smoothly, left, towards Timmy's room, as Deborah walks right towards hers.

'Did you see that?' I whisper.

'Yeah,' Lou says, calm as milk. 'That was a ghost light.'

It's confusion I feel first. Then a small swell of horror. This wasn't meant to be part of my story. Lou Gentile is supposed to be crazy, and we're supposed to laugh. He wasn't supposed to be *right*.

'It's probably best if you don't mention that to Deborah,' says Lou, scratching his belly underneath his black shirt.

And then I realise what it is. The cause of the tiny alarm in the diner. The reason that I've had this rising, unsettling feeling that something's not right with my story. It's the way Lou talks about his subject. He isn't telling me boastful tales of incredible bravery, or raging against imagined conspiracies or wearing a hat made out of tinfoil. This American eccentric, I suddenly realise, doesn't appear to be eccentric at all.

When Deborah returns, she senses immediately that something's happened. Cautiously, she asks what. There's an awkward pause, before her husband says, 'Nothing.'

We're now deep into the night's silent hours. Lou wants to see if he can get any 'Electronic Voice Phenomena' on his digital Dictaphone. He told me earlier that when there's pressure on the microphone, a little light on the recorder is triggered. 'It'll be dead quiet and it'll start to blink,' he said. 'How do you explain that? Sometimes I listen to the tapes on the way home and I have to turn it off. It's just about enough to scare the living bejesus out of you.'

I look at the grey Panasonic machine that's sitting on the coffee table and then at Lou, who's sitting forwards with his chin resting in his hands.

'Are there spirits in this house?' he says, evenly.

Nothing.

'Are there spirits in this house?'

The light begins to flash.

'Do you want this house removed from the land? Knock on the wall – once for yes, twice for no.'

I look up into the dark, stiffly, waiting. Nothing happens. Even though I don't believe in ghosts, when it becomes clear that there will be no knocking, a wave of relief washes through me.

It doesn't last for long. Lou picks the Dictaphone off the table and pulls the jog wheel back with his thumb. I then hear something that I never want to hear again, but will. It's like a cross between a dalek and a growling dog. It's cloaked in static as if it's coming from an epic distance away. And it sounds horribly like it's trying to say 'yes'.

As the EVP grumble on, we all hear one single, loud thump coming from upstairs. There's nobody there. I sit up rigid.

That's it, now. I admit it. I am scared. Really, very, very scared indeed. Deborah starts crying, softly. And then, something touches my back.

When I think back to this incident, it's my reaction that I remember most vividly. It wasn't: 'Ooh, did I just feel something?' It wasn't even: 'Hang on, I'm sure something just ... ' It was: 'Shit! Fuck! Something just fucking touched my fucking back!' and I leapt out of my chair. There was no momentarily unplaceable sensation, there was no doubt whatsoever. As far as I was concerned, in that instant, something had touched me, hard enough to launch me to my feet, definite enough to make me swear several times in front of mild-mannered Deborah and Tom. It was an instinctive, visceral reaction.

That night, back at my hotel, I try to get as much sleep as I can. What comes is fitful and thin. Fourteen hours later, Lou pulls up outside my hotel. We have another long drive ahead of us, this time to Stratford, New Jersey and the home of Kathy Ganiel.

'She's in serious need of help,' says Lou, as he indicates right and pulls out behind a taxi.

'Why is it down to you to help her, though?' I ask, reaching for the notebook in my back pocket. Lou never charges any money for his work, and all this staying up all night can't be good for his central heating engineering attention span.

Lou says, 'I grew up in a haunted house. I'd tell my parents and they thought I was nuts, you know? They sent me to a psychologist who said I was hallucinating. You know I'm not hallucinating.'

'What, exactly, weren't you hallucinating?'

'It started when I was about ten, always at about three in the morning. Black shadows would wake me up in the middle of the night, things would tap me on the leg, all the covers would be pulled off me, I mean real bizarre stuff, and I would tell this psychologist and she would say I'm having some dream-related stuff. But I'm not dreaming! I'm not dreaming! So I had nobody to talk to. I'm seeing psychologists, they're telling me I'm crazy, my parents are telling me I'm crazy. Until my sister and my brother started experiencing things, then I wasn't crazy any more. This was quite a few years later. We were sitting there – me and my wife, my sister and her boyfriend, my brother and his girlfriend – and were talking and there was the sound of people talking around us. Everybody looked at each other and ran out of the house and I'm the only one left sitting there, going, "I've heard all this before. Don't worry, that's all it's gonna do."'

By now, the night has returned completely. We're back on the freeway. The snow has stopped falling, but the landscape still looks petrified, buried and endless in the dark. I noticed, when Lou was talking, that the number three cropped up again. Last night, during a cigarette break, when I mentioned that the Carvens' clock had stopped at three, Lou asked me not to say anything for fear of frightening Deborah unnecessarily. Later, when he casually asked her about it, she'd said that the mechanism had snapped at exactly that position the other day, when she'd been winding it.

Sporadic flecks of snow have started dancing in front of the windscreen, and Lou, deep in thought, switches on the wipers. We pull out to overtake a couple of gigantic articulated lorries, a blazing rage of lights, smoke and noise. The traffic bombs around us and, it sounds like, through us, as we motor along in Lou's warm car.

'You know, when I was a kid,' he shouts above the freeway drone, 'I didn't have anyone there who could explain it. Maybe if I did, things could have turned out differently. Maybe I wouldn't have been so scared, you know? OK, sometimes I've had to go in there and tell people that what is going on is either psychological or natural phenomena. But, when there is a haunting, I'm there to help them make it stop, and so they don't think that they're crazy. I'm there so they don't commit suicide.'

'Really?' I say. 'Has that happened?'

'Yeah. That was a really sad case, in Johnstown, Pennsylvania. This guy starts telling me about these things that torment him in the middle of the night. And the next thing you know, he starts getting really sad and starts almost crying. He says, "You know, my wife died and I want to show you the letter she wrote." When I started reading this letter, I wanted to cry. It was just horrible. The lady goes ahead and tells her husband that she's killing herself because he doesn't believe that these things are tormenting her, and how she's going to take the gun and put it in her mouth and blow her brains out. So this guy starts telling me that he started experiencing the same stuff, that very night after she died. And he never believed her.' There's a small pause. 'Well, you've got to find these people some help. You'll see this lady tonight. If I was to walk away from her and never come back, she's gonna have these problems for the rest of her life.'

By the time we get to Kathy Ganiel's modest bungalow, the strange events of last night have been remoulded and rationalised by my brain. I've decided that I'd probably just been talked up into a state of wild and paranoid fear. Tonight, I'm determined to be on my guard. I am a professional journalist who deals only in hard facts. I will not start believing in the impossible, just because somebody's turned all the lights off and told me a story about a spoiled toddler who's managed to get away with blaming his untidy bedroom on the dead.

'This lady is basically possessed,' says Lou, quietly, as we unload his demonology kit from the boot of his car. 'She's used a Ouija board, and other things. This is stuff that you do not play around with.'

Kathy doesn't look possessed. She's a petite, pale and bird-like 38-year-old mother and wife. Her compact home is clean and comfortable, if basic. Tonight, her husband is working and her kids are long asleep. She's preparing tea in the kitchen, which is cluttered with baby bottles, fridge magnets, coloured toys and drying dishes. As the washing machine goes into a spin cycle,

Lou, still trudging back and forth with his boxes and cables, pops his head round the corner and prompts her to tell me her story.

'Well,' she says shyly, stirring the drinks, 'my dad died five years ago and I started dabbling with a Ouija board. At first it didn't work, but then it gave me information about stuff that happened with my great-grandparents, and I'd go check with my uncle and he'd say, "Yeah, that's what happened." So,' she says, handing me a mug of milky tea, 'I was in bed one night and I felt like someone was watching me. And I'm not usually paranoid. I had the covers over my head but I managed to peek and there was a black shadow standing at the side of me. Then the foot of my bed starts shaking. And I'd get scratching noises and knocking and whispering. I would see lights. And this would happen every night. Sometimes I would hear my name, I would distinctly hear "Kathy". Then the children started to hear stuff and I played it down. At the time, I didn't think it was evil. I thought it was my dad.' She pauses briefly and fiddles, distractedly, with her wedding ring. 'But then I would think, why would my dad come at night and shake my bed? Why would he stand over me and frighten me?'

'Do you believe in God?' I ask. Before Kathy can answer Lou appears and interrupts her from the dining area, where he's setting up a large laptop. There's a look on Lou's face. And it's directed at me.

'She keeps using divination to try and contact the dead,' he says. 'She uses a necklace. She holds the top of it between her fingers so it acts like a pendulum, and then she asks questions. If it sways back and forth, that's a yes. If it swings in a circle, that's a no. And when she's not doing that, she's using a Ouija board.'

'When did you last use the Ouija board?' I ask Kathy.

'About three weeks ago,' she whispers.

'She would tell me she'd stopped,' says Lou, 'but ... '

'It becomes an addiction,' Kathy says, looking at the floor. 'I'm trying very hard.'

'We've had a priest over here,' Lou says, as he waits for his PC to boot up.

'Lou said I did some inappropriate behaviour. I don't recall.' She looks into the middle distance, puzzled. 'I do remember that I wanted to smack him in the face, and I would never do a thing like that. I don't remember ... ' She pauses and chews on her bottom lip, frowning. 'I think I wanted to take a cigarette and stick him with it ... and he was fairly nice ... Lou?' she asks, looking up. 'What did I say to him?'

'I don't really want to go into that,' he says, walking off again to fiddle with a tripod.

It's just after 1 a.m. Lou has the camera trained on the kitchen and we're in the living-room area of the L-shaped, open-plan part of the bungalow. There are small, silver-framed baby pictures on the shelves, in amongst crucifixes and small pots of miscellaneous family detritus – badges, safety pins and half-used match-books. A wall-clock ticks in the background, as we sit on the old, slightly grubby three-piece suite – me on the armchair, Lou and Kathy on the sofa. The demonologist sits forward to click the monitor on. Instantly, I want to go home.

There are ghost lights in Kathy's kitchen. Lots of them. Every few seconds another bright globule appears and travels through the air for a few seconds before disappearing. Larger, static discs throb in and throb out again. It's like nothing I've ever seen before, like nothing I could ever have imagined.

'Does anybody want more tea?' Kathy asks.

'Sure,' says Lou.

Kathy walks around the corner and we watch her appear on the monitor.

'Wow,' I manage to say. 'Christ.'

'Don't say that in front of Kathy,' Lou whispers. 'You already did that once. I told you. If you provoke this lady with religious language, she'll start to go under.'

'What happens when she "goes under"?' I ask, watching the image of Kathy filling the kettle on the screen while the sound of the rushing water comes from the kitchen. Above her head and around her, globules float and dart and throb.

'Her eyes and forehead will go down. It's like she's a completely different person. Psychologically, she's been tested. She's not bi-polar, schizophrenic or any of the other stuff. She might not look like she'd harm anybody, but believe me, when she goes into full possession, she can do anything. You have to be very, very careful. When the priest was here she really should've been restrained.'

'What was she saying?' I whisper.

Lou leans towards me and speaks quietly. 'Something along the lines of, "Take your salt and stick it up your ass, your Christ can't help me." Something like that. "Get your fucking salt and stick it up your fucking ... "'

At that moment, Kathy comes back with the drinks. We sit calmly for a time. The wooden squeaks and taps of the house play lightly on top of the slushy sound of the cars on the still, suburban road outside. Lou leans forward. Tonight, his Dictaphone is in the kitchen. We can see it on the monitor, standing next to a packet of Pop Tarts.

'Is there a spirit in the house with me now?'

We watch the recorder's LED start flashing in a hectic and terrifying silence. I cough loudly three times to clear my throat and watch the magic, quietly aghast.

'Christ ... ' I mutter.

'Show me a sign in the kitchen that you're with us now,' says Lou.

Slowly, I become aware that Kathy is staring at me. This is it, I think. It's happened. I can feel it, the air is thick with it, tense and electric and snarling. Kathy has gone under. I try to pretend it isn't happening. I stay fixed on the green screen and watch the LED flicker and fade. Then, Kathy starts mumbling gruffly. The only word I can make out is 'power'.

'What was that, Kathy?' says Lou.

She sits up, bolt still and then, in a heavy voice, says, 'Foolish men sit around and wait for displays of power.'

'You OK, Kathy?' Lou says.

'Playing with fire.'

I stare at the lights moving and throbbing on the screen, and try to block out the fact that Kathy's eyes are boring into my head.

'Do you want to see a trick?'

I don't move.

'No, Kathy, Will does not want to see a trick.'

'Why not?'

'Because you're going to do some divination.'

'Do you want to see a trick?'

'Is that a chain you've got there, Kathy?'

I try to look round with only my right eye. Kathy has the tips of two fingers in the coin pocket of her tight jeans. I can see that, inside it, she is fiddling with a fine gold necklace, the one I assume that she uses to contact the dead. There's a long, heavily freighted silence. Then, she stands up and walks out.

I watch her go, then breathe out, my shoulders collapsing. I look around, checking that the front door is still in the same place that it was, while Lou goes into the kitchen to retrieve his Dictaphone. He plays his recording back. Compared to last night, the barks and growls in this place are diabolical and furious.

'It just goes right through you,' Lou says, and I notice that the sound of my coughing is present and in the correct place on the recording. It's 3.22 a.m.

'The best chance of seeing a ghost is when you're by yourself,' Lou whispers. I look over to his face in the light of the glowing monitor. He looks like a domesticated and slightly evil Elvis.

'Some people say that this stuff is just your innermost fears coming to a conscious reality. But there are tons of cases where many people have seen the exact same thing. I'll stay here, you go into the kitchen for a while and tell me what happens.'

Obediently and condemned, I walk around the corner. And I'm frightened. It's as if the dark is thick and airless and I have to push my way through it. Eventually, I find a low chair in the corner and sit down, amongst discarded toys and books and ironing.

For a while, everything is normal. The settling house pocks, ticks and knocks. The cars swoosh past and my breathing rises and falls like a breaking sea. And then I see a streak of light dart across round by my feet. Then, a rapid series of knocks bang the wall. I know I've never been this scared before, because I've never felt the sensation of thousands of needle pricks coursing up my legs and through my torso. Suddenly, I am furious that I've put myself in this situation. I am, literally, too frightened to move my eyeballs. I sit hard and still, chest up and jaw stopped. I'm about to break every single one of Lou's rules except the one about laughing.

Five minutes later, Lou switches on the lights. I look, searching, into the bright, open air, in the places that the ghost lights appeared on the screen. Nothing. I tell Lou about the knocking and the lights. He shrugs and says, simply, 'OK.'

So what happened in Pennsylvania that winter? Whatever it was, it prickled at me for a long time afterwards. I became obsessed with what I'd seen, as did other people. Friends of mine told friends of theirs about Lou Gentile and his dark vocation. One afternoon, I got a panicked phone call from somebody who knew somebody who knew somebody I knew. She wanted to know exactly what Lou had told me about the number three. She said that she and her father used to wake up at the same time every night. It became a running joke in the family. Neither of them knew what it was that interrupted their sleep at 3.33 a.m. every morning, but now, after hearing my story, she was frightened.

I know what I saw, and it wasn't, as one parapsychologist has told me since, 'insects'. And I'm convinced that I wasn't taken in by some conspiracy or hoax. You could argue that Kathy was mentally ill, but that doesn't explain the knocking or the ghost lights. If they were just specks of household dust (or, indeed, insects), how did the things appear and disappear in mid-air?

And why did they behave so strangely? What about the streak of light? And the knocking? What about the EVP? My coughing proves it wasn't pre-recorded and the sound was unlikely to have been some sort of static interference because it came in response to Lou's questions. And interference doesn't answer questions.

I also wonder if the Carvens could have made their ghost stories up in order to sue the property developer who sold them tainted land. But would they really risk all that ridicule? Why would they perjure themselves in such a ludicrous fashion? And, anyway, what court would rule, legally, that ghosts exist? Aside from all of that, what about the evidence that Lou found? The EVP and the ghost light that appeared right on cue? How do you explain the fact that something touched me?

This is my problem: if I accept that ghosts do exist, then the hard walls of my straightforward and rational world fall down like colossal reality dominos. Because if we don't die when we die, then nothing is as it seems and everything is up for questioning. All logic is gone. The priests, with all their smoke, spells and bad news, could turn out to have been right after all. There could be an afterlife, and if there is, that means there could be angels and demons and heaven and hell and rules of right and wrong by which I should be living. But, there again, they could, quite simply, just be insects.

So, I need to find some answers. Because, right now, I'm haunted by questions.

To: Lou Gentile
From: Will Storr
Subject: Ghosts

Hi Lou

Following my trip to Philadelphia, I have decided to spend some time looking into the subject further. I'd really appreciate it if you could keep in touch over the next few months and let me know what's going on with you. Give me a shout if anything really mental happens.

Thanks.
Will

1

'Are you Sir Thomas Sackville?'

He's a man's man, is Christopher Tuckett. Rosy-cheeked and countryside-fit, rugged-faced and handsome. The 34-year-old assistant property manager (events) of Michelham Priory is all thick-cord trousers, rolled-up shirt-sleeves and shooting the shit out of pheasants. That's what the (events) part of his job title means, by the way – countryside pursuits. He adores them. And they're his job. So, you believe him when he says, 'I'm not the type who's easily fobbed when it comes to spooks and whatnot.' (I believe him, anyway: I've seen a framed photograph in the lobby of him with a peregrine falcon perched on his arm. He's giving it a steely look.)

'I was a right cynic,' Christopher continues, leaning back on a thick stone ledge. 'I didn't believe in ghosts. Still don't. I'll argue the toss about anything.'

'What,' I say, 'even when you've got a tornado in your bedroom?'

But the rest of the Ghost Club don't laugh. They just stand there and blink back at him.

It would, perhaps, have been better for me if the Ghost Club had turned out to be a support group, the sort that sits in circles in meeting halls, a cuddly community of crumpled victims who wear name badges, have tearful, confessional moments and a twelve-step rationality-recovery programme. Unfortunately, however, they're not. They're unashamed and unreconstructed Ghostaholics. And I am their newest member.

A couple of weeks ago, I decided that my only option is to confront my fears, to charge bravely, head first, onwards towards the answers. So, I tracked down the Ghost Club's website, printed off an application form, filled it in and popped it in the post.

Barely two weeks later, I am here in Michelham, East Sussex on my first investigation. The thirteenth-century priory looms in misty, silent grounds, surrounded by a still-treacherous moat. Its high sloping roof, tough Tudor chimneys and thin, suspicious windows give the building the air of a defensive, growling animal. It doesn't want you anywhere near it.

We're standing in the undercroft, the priory's large, square, stone-walled entrance room. It has a low roof of curved arches that honeycomb across the ceiling and gather together into a large column that comes down into the middle of the room. It's 5:17 a.m.

'Well,' says Christopher, rubbing his chin, as if he's being made to consider the tornado incident properly for the first time. 'Actually, I would say it was more like a mini-tornado. Me and my wife used to hear it in the kitchen. Then it would come down the corridor and into the bedroom. It would be there for a good ... two or three minutes?'

'And it would ... what?' I ask. 'Blow stuff around?'

'Mmm, yeah,' he says, nodding with his hand still cradling his jaw. 'The curtains would be flat on the ceiling. Then you'd hear it go back up the corridor and into the kitchen.' He ponders the mini-tornado for a few more moments before muttering, 'That would wake you up.'

In the corner of the undercroft, a life-size waxwork model of an Augustinian canon gazes at us piously from beneath a dark cloth hood.

Christopher has permitted the Ghost Club to investigate the property that he's allowed to live in as an employee of Sussex Past, the site's owner. He's come down to lock up behind us.

'My wife left about a year ago,' he says, folding his beefy, bramble-grazed forearms. 'She couldn't handle it here. I had to make a choice between my marriage and my job. And the job's quite good.'

The worst time for Christopher, he confides, was when he first moved in. He wasn't allowed to live with Sue until they were married, so he had to spend two weeks living here alone.

'The flat upstairs is quite huge,' he says, 'big rooms. And there was no furniture up there except one circular table and a bureau in the kitchen, which we decided to move into the sitting room. I remember waking up at three in the morning and you could hear this thing moving across the floor, as clear as you like. And you knew what it was straight away. It was the bureau, on these small brass caster wheels. I remember thinking, oh dear. Well, this noise seemed like it went on for ever. But it stopped and, well, eventually curiosity gets the better of you. I went into the sitting room and turned the light on. The bureau had been pushed up against the corner and it looked like it had only come out that far,' he says, measuring about a yard with his hands, 'but when you looked at the floor, there was this huge figure of eight scratch. It had scored all the polish from the floor, where one of the wheels had seized up.'

'Christ,' I say, involuntarily.

'Now, I have to admit I do sometimes find it a little difficult to relax. Some nights it's busier than others. You always know when it's going to kick off. You know when you go into a pub and everyone stops and looks at you? That's the feeling. Sometimes I'll walk in, get a chill down my spine and I'll go flat out up the stairs. But I just think, psychologically, that's me snowballing in my mind.'

I know what he means. I've done a bit of psychological snowballing in my own mind over the last few hours.

It started in the gatehouse, a centuries-old castle-like tower that guards the entrance to the priory over the moat. I was with a senior member of the Ghost Club called Lance, and a couple in their early twenties called Natalie and Dane. I'd gravitated towards Lance early on because he looked like a man who knew what he was talking about. Male-pattern-bald and dressed down in a black T-shirt with studious wire-rimmed glasses and a cautious and precise way with words, Lance brings to mind a very kind, curious and learned old mole. I followed him about earlier as he set up various ghost traps. He sprinkled talc on a bookcase to capture phantom handprints; he put a teddy on top of the stairs and marked its exact location in case a recently reported apparition of a little girl is coaxed out of the shadows to play with it; and he talked about EMF and gaussmeters, which measure disturbances in the magnetic field that have been shown to take place where there's ghostly activity.

'The Ghost Club's stated aim is to observe, record and monitor possible phenomena,' he says. 'We're not ghosthunters in the sense of stamp collectors. We're investigators. It's a very imprecise science and our purpose is to gather evidence and to try and improve the methods by which we do so.'

When we reach the gatehouse I notice, through the half-light, that Lance is carrying two thin metal L-shaped poles.

'Dowsing rods,' he says, noting my noticing. 'You put them in your hands like this.' He curls his fingers and rests one rod in each hand, so the longer part sticks out from the top at right angles. 'Ask a question, and if the rods cross, that means yes.'

'What, they just cross?' I ask. 'Just like that, of their own accord?'

'It works with about 80 per cent of people,' he says, 'but nobody quite knows how. A simple test is to approach somebody with a pair of rods and, in almost all cases, they will part as you get near them. There is no physical contact, but you're producing physical results.'

'Can I hold them?' I ask.

I take the rods in my palm and weigh them. They're strong, but incredibly light. I'm astonished. If what Lance is telling me is true, I am holding two bona fide mysteries of the paranormal in my left hand. Could these rods really cross of their own accord? And on cue, when you asked them a question? Even if they move aside when you approach someone – that's scientifically absurd, isn't it? Physical results without physical contact? That breaks the laws of physics.

I carefully pass them back to Lance. I don't want to break them. They must be valuable. Where do you get hold of something like that? I wonder.

'They're Sketchley's coat-hangers,' Lance says, 'bent into shape.'

Behind Lance, a man in a leather coat bends down and unclicks a large metal briefcase. He's got a milky complexion, close-cropped hair and wide, pale eyes that look as if they're in a permanent state of examining things. He opens the lid of his case and pulls out two enormous dowsing rods, like pistols. They're bent into perfect right angles and, even in this dull light, they manage to flicker and glint. I walk up to him and introduce myself. He shakes my hand and gives me a business card – 'Paolo Summat, Paranormal Investigator'.

I feel disappointed when I'm told that Paolo is not in my group. He's been assigned to work with a woman called Anne. Anne makes me nervous. She walks around the place silently and on her own. She has no understandable expression and always keeps her arms hidden under her long coat. She doesn't spook Paolo, though. I watch them walk off, down a gloomy corridor, while Lance, Natalie, Dane and I leave the undercroft and the main building to walk, by torchlight, across the lawns towards the gatehouse.

The inside of this large, dusty, 800-year-old room, which sits directly over the priory's entrance, is so black that you can feel it pushing against your eyeballs. Sitting here, against the cold stone, the only thing I can make out are two thin arrow slits in the wall. The night sky outside them is tinted and glowing with cloudy moonlight. Suddenly, there's a flash. It's Dane's digital camera. I can see his face as he studies the screen on the back of it. He's looking for orbs – or 'ghost lights', as Lou Gentile would have it. He flashes and checks, flashes and checks. The thought that orbs could be invisibly zipping about the place makes me uneasy and I shift about, nervously.

'I've got one over by the fireplace,' Dane says. In the torchlight that his girlfriend is shining on him, I watch Dane go and kneel down on the dusty floorboards where he saw the orb. He has a set of rods in his hands.

'Is there a spirit present with us now?' begins Lance.

The rods don't move.

'If there is a spirit present, will you please cross the rods?' he says.

One of the rods starts to shake, tentatively. I squint and try to focus to see if Dane is moving it.

'It's struggling a bit,' Dane says. 'Do you want a go, Will?'

'Yes,' says Lance, 'why don't you give it a try?'

After Lou's warnings about divination, I was hoping this wasn't going to happen. Nervously, in the feeble torchlight, I walk over and sit down in the jaws of the echoey medieval fireplace.

'Is there a spirit present?' Lance says.

I stare at the rods. I decide to will them, to plead with them, to beg them. Inside my head, I shout *no.*

'If there is a spirit present,' says Lance, 'will you please cross the rods?'

No, don't, please, I think as loudly as I can.

'If there is a spirit present,' repeats Lance, 'will you please cross the rods?'

At first I feel a bit guilty. Then I feel a bit bored.

All right, I say silently, *cross.*

I shift the weight between my legs again and look up and around me, for some sort of visual reassurance that the world is still as I know it. All I can see are the thin, glowing arrow slits. They look as though they're floating in space.

'If there's a spirit present,' says Lance, 'will you please cross the rods?'

And then, to my rising horror, the left rod starts to tremble.

'Please cross the rods,' says Lance.

Slowly, shakily, as if it's being tugged by an invisible moth, the left rod crosses over the right one, which stays entirely static. My hands, I could swear, do not move.

A flash.

'Got an orb, right under your chin, Will,' says Natalie, her nose buried in the back of her digital camera.

'Does anybody else want to try?' I say, getting up and swatting the empty air under my chin with my hand.

Feeling the need to gather myself back together somewhere I'm allowed to put the lights on, I walk back to the undercroft for a think.

As I enter the room I notice a hard plastic briefcase with its lid open. Inside is another pair of dowsing rods. I decide to take them out and test them. Up in the gatehouse, they crossed only when I instructed them to. So maybe, I think, my hands are moving subconsciously. There could be tiny muscle twitches attached to my thoughts, which are making the thin metal poles move in minute, juddering fractions.

I hook the rods into my curled fingers, stare at them for a time and, when I'm ready, unleash a torrent of silent will. *Cross!* I demand with my mind's most mountainous voice. *Cross!* And, just like in the gatehouse, they twitch into life.

I put the rods down. Mystery solved. Unless I'm to believe that an invisible spirit heard my wish for the rods to cross and decided to give them a little nudge in the right direction, it would seem that this divination thing is all a big misunderstanding. I think about Kathy and the gold necklace that she'd hang from between her fingertips. Was she just talking to herself all that time?

Just then, Lance walks in with Natalie and Dane. They noisily beat the cold out of their clothes and Lance tells me that they had a short conversation, via the rods, with a spirit who claimed to be a Benedictine monk. 'Even though,' says Lance, curiously, 'this was an Augustinian priory.'

I tell Lance about Lou Gentile's uncompromising warning

about divination. Does he agree that, in using dowsing rods, we might be dangerously meddling in dark sports?

'It's a difficult one, that,' he says as he sits down next to me. 'I, personally, would not use a Ouija board. I can't explain why I draw that line other than the fact that, yes, you are in more direct contact with a spirit.'

'Can we be sure it's a spirit, though?' I ask, and I tell him about my experiment with the rods.

'Well, there is some evidence that using a pendulum is picking up on small motor movements generated by the subconscious, and in the same way, I suppose divining rods could be picking up on ... well, yes, a presence, but not actually a possessive one.'

'Have you ever been scared?' I ask him.

'No,' he says immediately, 'I haven't. The closest thing I've come to being scared is at a pub in Buckinghamshire where we heard a strange snuffling sound.'

'What was it?' I ask.

'It turned out to be a hedgehog.'

I look at Lance. 'So in all these years, in all these haunted places, you've never once been frightened?'

'It's possibly because I'm too mundane,' he says, with a sad shrug. 'I don't think I pose a threat to a ghost. The only one I've ever seen is my cat, Emily. I saw her in my kitchen when she was at the vet's, being put down. That wasn't full of signs and portents, but I know what I saw and I saw that cat.'

Lance tells me that there are four main types of ghost. There are replay hauntings – disembodied footsteps on stairs or white ladies in lakes that just do what they do over and over again. There are crisis apparitions – visions of loved ones mysteriously appearing at the time of their death. Then, there are ghosts with a purpose, who come back in order to deal with business left unfinished during their lifetime. And then there are polts.

'Polts?'

'Poltergeists. They seem to be people-centred.'

'How do you think I should go about finding out about poltergeists?' I ask.

'Well, obviously, you should see if Maurice Grosse is willing to talk to you,' he says.

'Maurice who?'

'You've not heard of Maurice Grosse?' says Lance, his curiosity almost popping. 'Well, I strongly recommend that you interview him. He's in his eighties now, a scientist by background and probably our most distinguished living investigator of polts. Have you heard of the Enfield poltergeist case?'

'Yes,' I say, 'I have. A demonologist told me about it.'

'Well, Maurice rigorously investigated that case, and witnessed it first-hand. You know,' Lance says, settling into his subject, 'it's always occurred to me that, given nature works within an economic framework, it doesn't seem economic or reasonable that people spend their lives gaining wisdom and having the rough edges knocked off them and then, just because of a physical flaw, that mental progress is lost.'

No, I think, I suppose it doesn't. What's the point of life if it just ends? Why would nature do that to itself?

'Another one of the interesting things about ghosts is that I can't think of any other phenomenon which is so consistent across time, class and cultures. The surveys the Society for Psychical Research – the SPR – carried out in the 1880s were finding that a third of the population believed in ghosts, about a third were agnostic and a third were strongly sceptical. That was 125 years ago, and it hasn't changed much. I've been on courses with senior civil servants and those proportions hold good there. I just find that very fascinating.'

Whilst he's talking, it strikes me that Lance is one of the most exactingly rational men that I have ever met. In fact, his rationality is his defining characteristic. It's evident in every word that he says, every expression on his face, every meticulous note that he makes in his Ghost Club Event Log. And yet, many people would ridicule his fascination with the paranormal. They'd listen to his tale about Emily the ghost cat, and wink and whisper and snigger out of shot. They'd say he was mental, and when I'd insist that he wasn't, they'd conclude that he was lying.

Usually, when I talk about my experiences with Lou Gentile, sceptical types tell me that I was hoaxed. But I don't buy that. And not only because I was impressed with the witnesses and the evidence. Lou Gentile, I'm sure, is convinced of the truth of what he says.

As are the people I've met tonight. Why would Tuckett invent stories about a weather system in his bedroom? Bearing in mind that Michelham doesn't promote itself as a 'haunted' tourist attraction, and that Tuckett still stubbornly refuses to believe in ghosts. Why would Lance try to bamboozle me with a yarn about Emily, the spectral cat?

I don't believe that the world is full of cunning and convincing tricksters who are armed with rewired tape recorders, smoke machines, apparition projectors and hidden tornado turbines. It just doesn't add up that, for generations, thousands of clever people have spent hours concocting tales and elaborate practical tricks to fool people just for ... for what?

So, I have made a decision. Unless I have good reason to think that I am being given the sly one, I am always going to assume that the teller believes that they are telling me the truth. I don't believe that the people I am going to be meeting – people who spend their spare time obsessing about and fiddling with the paranormal – would bother. And secondly, I just think I'll be able to tell if somebody's trying to cock me a nonsense.

So, if the people that I've met so far aren't lying, then it means there might have been demonic faces and black shadows and shaking beds and an invisible force fucking with Christopher's furniture. And if that lot's been happening, I have to find out why and how because, really, the implications for me are stark and they are deafening. Because if there is something lurking behind the everyday, if there are hidden rules and dimensions to existence, then perhaps, I think, my days need to be lived more meticulously. Lou and the priests would have me believe that if I behave badly, I'm going to end up in hell. Before Philadelphia, I would have dismissed all that as superstitious nonsense. But, since then, my faith in the nature of nonsense has been radically shaken. And if they're right, if there is an afterlife, and my actions now will have an impact on me for all eternity, then I really do need to know. And sharpish.

I look up. I am alone in the undercroft at just gone 3 a.m. After having found a very human answer to the divination question, I feel brave enough to have a wander around on my own.

As the winter sun has sunk, this place has gradually come to feel less like a polished tourist attraction and more like an extremely old house with a few bits of red rope draped about the place. Beyond the priory's moat are ancient fields. They are small, with irregular boundaries marked by sinewy hedges, and they're dotted with oak trees so old they appear to have developed their own personalities. The roads we drove in on are the same winding tracks that have linked the houses, farms and the outlying villages of Herstmonceaux, Horsebridge and Abbot's Wood for many long and troubled centuries. They're thin, ditch-lined and unlit and empty at night.

Inside the priory, as I walk around, are wood-panelled walls, exposed beams and austere paintings of grave men with hooded eyes and high-buttoned collars. These are the faces of Sir Thomas Sackville, John Foote, William Child and James Eglington Anderson Gwynne, the men who once inhabited these rooms, who ate roast pig and argued over these tables and regarded themselves in these mirrors and slept and died in these beds. It could be these faces that keep Christopher Tuckett awake in the night.

I creep up a wide set of stairs, turn a corner and pad past a 'Michelham Through the Ages' display. There are three more stairs and beyond that a large room. I walk in and, instantly, I'm spooked. Perhaps it's the size of the space, the deep shadowed corners and looming, huge furniture. Or perhaps it's the silence that the night seems to amplify. I take a few steps forward, as gently as possible, scared I might somehow provoke something lurking in the quiet. To my right are lead-latticed windows, which look out onto the shushing trees and lawns. Silhouettes of antique chests and cabinets line the walls guiltily, like a caught-in-the-act gang trying to keep undetected in the dark. I walk forwards. In front of me, I notice a beautiful oak chest. One of its drawers is sticking out. As the rest of the building is faultlessly immaculate, this strikes me as strange. I decide to pull the dowsing rods out of my back pocket for an idle test. I hook them over my fingers and walk towards the drawer. I take one step – nothing. Another step – nothing. Another step – nothing. And then, just as the rods cover the open drawer, they tear across each other. One of them actually spins right round and stops only when it touches my jumper. It was nothing like what happened in the gatehouse. It had real, independent force.

'Shit!' I shout and pelt back down the stairs.

I shouldn't be this jumpy, I think, crossly, as I sit in the safe, bright undercroft and dry my palms on my trousers. It appears that I might have dismissed dowsing a little hastily.

Ten minutes later, I've managed to persuade Lance and Paolo away from examining a cold spot of air in the kitchen. They've agreed to offer me a second opinion. We enter the room behind Paolo, whose long rods are poised, sniffing the dark air like the feelers of a giant robot ant.

'There's something in here,' says Paolo, 'there's definitely something in here.'

He creeps forward with Lance and me two steps behind.

'What is it?' I ask.

Paolo remains silent. He stops beside a velvet-covered chair.

'I can sense you are standing right behind me,' he says to the room. 'If that's the case, would you please cross the rods.'

'I think it's trying to cross slightly,' says Lance. 'It's now crossing.'

'OK,' says Paolo to the spirit. 'Thank you. Your presence is very strong and we need to talk to you. Is there anything you would like to say to us?'

'I feel sick,' I say.

'Yeah,' says Paolo, 'you know when it's a yes because you feel a surge.'

'What do you think the surge means?' I say.

'I've got a feeling it's "fuck off", actually. Oh,' says Paolo, suddenly looking even more pale, 'I find it so hard to do questions. Lance, would you talk to it?'

'Did you live here?' says Lance. 'Were you a resident of Michelham Priory? If so, please cross the rods.'

Again, they pull together over Paolo's hands in a grand, swooping movement.

'Are you feeling a real rage, yeah?' Paolo asks me, 'like it really wants to have a go at us?'

'I can feel it up my back,' I say.

'Did you die here at Michelham? Are you dead? If you're dead, could you please cross the rods?'

Lance and I are standing close to Paolo at the edge of the empty room. We're all staring at the rods, which are illuminated by the moonlight streaming in through one of the thin windows. They quiver slightly with Paolo's heartbeat. And then they cross. He shakes them back to open with a well-practised flick.

'Are you angry with us?' says Lance.

The rods swing together again.

'He's livid with us, yeah,' Paolo says. 'If you give us ten minutes of your time, we'll leave you in peace. Would that be OK?'

They cross again and Paolo resets them, quickly.

'Would somebody please make note of the time? I'm going to have to keep my promise.'

'It's three forty-one,' says Lance. He clears his throat and addresses the ghost. 'Are you Sir Thomas Sackville O' Dorset? Chancellor of the University of Oxford? Are you that true and puissant knight?''

'Look at that,' whispers Paolo, as they cross again. 'It's right-handed, this ghost.'

'I'm getting weird sensations in my head, twitching and that,' I say.

'Yeah, I'm getting that,' says Paolo. 'It wants to smack us, basically. It's saying, "How dare you?"'

'Sir Thomas,' says Lance, 'how many other beings whom we cannot see are here with us? Is it three other beings in the building with us? … Is it four other beings in the building with us? … Is it two other beings in the building with us? Thank you, Sir Thomas. Is one of those beings a little girl? Thank you, Sir Thomas. Was it you who opened the drawer to Paolo's left?'

'I feel really sick,' Paolo says, his rods drooping, 'I want to get out of this room. I can't cope any more.'

'God,' I say, backing towards the door, 'have we really fucked Sir Thomas off?'

'Yes,' says Lance, as we briskly exit, 'and almost certainly the last person to do that was Queen Elizabeth I.'

To: Tim Laverty
From: Will Storr
Subject: Michelham Priory

Hi Tim

I wonder if you can help me. I am currently doing some research into ghosts. I was on the Internet doing some research about Michelham Priory and I came across your appeal for information in a chat-room, following the events that took place during a school outing at the priory. I was wondering if you could fill me in on exactly what happened?

Many thanks.
Will Storr

To: Will Storr
From: Tim Laverty
Subject: Michelham Priory

Hello Will,

Thanks for your email. Of course I'd be happy to let you know what happened.

I'm a member of teaching staff at a secondary school near Lewes in Sussex. Last Wednesday we took a bunch of students out to Michelham Priory. All in all, it was quite an odd school trip. A number of kids reported paranormal phenomena, including:

- Cold spots (at the foot of the stairs and in the Prior's Chamber).
- Stones being dropped onto the centre of the floor in one of the upper rooms, from no apparent source (I witnessed this, too).
- A very distinct noise of sneezing from the empty nursery.
- Sudden feelings of nausea in the top room at the gatehouse, which disappeared when they moved from the room.
- A kid being tapped on the shoulder by an invisible source in the barn (he nearly leapt out of his skin).

• Another kid hearing the names Mary, Elizabeth and Rosemary spoken in the nursery (no one else heard this, but he was insistent).

I brought along a Dictaphone, with which I recorded the tour on the upper floors, and on playing back the cassette tape immediately after the tour, there are three shouts of either "No" or "Whoa" recorded in the Prior's Chamber, and what sounds like five screams, possibly that of a young girl, recorded in the upstairs music room, adjoining the nursery.

Of all this, it's the tape recordings which seem to be the most significant. The kids, in small groups, were all remarkably well behaved and there was no shouting or screaming going on in the room at the time, and this was confirmed by the tour guide and other staff members. The recordings are quite disturbing; staff who asked to listen to them later were visibly shaken. I've made a rough copy for the school to use, and the Priory asked for a copy which can be added to their archives. Hope that's of some use.

Tim Laverty

2

'It's not all coming from the trees'

What if I've been wrong about everything? What if everybody's lives aren't, as I'd thought, like eighty-year-long films, where everyone plays their own hero and fades to black in the end? What's looking increasingly unlikely to me is the idea that the only world that exists is the one that we can see in front of us. Because there have now been two occasions when I've set out to find ghostly phenomena. And both times, I have found it. That's a 100 per cent hit rate.

Where, I keep wondering, will this search lead me? Could it be taking me, in a very circuitous route, back to church? I keep getting a flashback to a three-second sliver of an event that happened a few years ago. I was having a meal at my parents' house when my mum proudly passed comment about how spiritual her family was. Quietly, I murmured between mouthfuls that I wasn't spiritual. Not at all. Even when I've been lost in my darkest, most desolate acres, I've never felt the presence of God or even been tempted to look for it. But Mother knows best. So when I said it – *I'm not spiritual* – she gave me a self-assured, beatific smile and announced, 'Oh you will be. You will be.'

At the time, at the table, I was silently furious. But maybe she was right. Maybe this is the beginning of a spiritual journey, one that will kidnap my rational self and drag it off to church for confession. Because, for a lapsed Catholic like me, you can't just decide to believe in ghosts and leave it at that. Because if ghosts exist, and they have intelligence, then the ramifications are vast and they are terrifying. Because it means there is an afterlife. And that is hard evidence that I could have been rash in my lapsing.

The afterlife – the concept of heaven and hell – is at the very core of Christianity. The fear of damnation and lust for paradise is the twin-valve engine that drives the worldwide faithful. And, put simply, there either is an afterlife or there isn't. To me, deciding that the afterlife does exist might be the most important decision I ever make. Because if heaven is a reality, and entry depends upon how you live your life – which becomes, in effect, one long moral assault course – then I need to start doing some exercise.

But it's a little more complicated than that. Christians do believe in an afterlife, and ghosts may well turn out to be proof of that. But they also believe in the devil. And every good priest would agree with Lou Gentile that Kathy Ganiel wasn't being haunted by any human spirit, but by a demon sent by Satan. A demon she opened her door to when she made the fearsome mistake of using divination.

It was Father Bill who first warned me about Beelzebub. I was thirteen, and had been caught bunking off Sunday school to do a Ouija board in a multi-storey car park with some friends. As part of my punishment, it was arranged that I would have a sit-down telling-off from Father Bill, the local priest who, I was told, was the parish's appointed exorcist.

He sat me down in one of the gloomy flats that housed the clergy and warned me to stay away from the world of the occult. Many boys dabble, he said, only to be struck down in the future by hell-borne problems. And it's not just obvious things, he said. It's not all possessions and hauntings and black shadows by the bed. Satan can come knocking wearing a more mundane coat. The Ouija abuser could have health or personal problems, or their luck could just turn rotten. Most often, the afflicted simply find that their faith in God mysteriously drains away. The invisible world of the undead, the world of ghosts and spirits, is the world where the devil lives, he told me calmly. And if you go looking for the devil, the devil will find you.

I've never had cause to think much about the strange and menacing warning that Father Bill gave me all those years ago. But, as this journey takes me closer to the tumble-down world of the mystics and the priests, and my rationality slowly dissolves, like scientific sugar lumps in supernatural tea, I'm becoming increasingly worried about my spiritual safety. Father Bill warned me about the dark side, and, suddenly, I can feel its weird shadow falling across me. So it's obvious what I need next – some advice from a druid.

For that, I've travelled up to a chewing-gum-dappled brick suburbia, half an hour north of Newcastle, to meet Stephen. I've heard that he teaches beginners how to keep themselves safe from supernatural harm in vigil situations. So far, the only anti-ghost defence technique I've been shown is Lou's 'white Christ light'. And, as that relies solely on my unshakeable faith, I thought it would be a good idea to load myself up with as many other ones as possible. If I'm going to spend my evenings and weekends hanging about in places where the structure of time, logic and death is unstable, I'm probably going to need the protection of some sort of psychic hard-hat.

'The mind can often see what was then as well as what is now,' Stephen says, with his eyes set to enigmatic, when I meet him. It's half past eleven on a Saturday morning and we're standing in the middle of a neat, grassy family park, over the road from a Texaco petrol station and a hairdresser's called Hot Cutz. The sky is mercurial and sagging. There's a freezing, acid wind and ticks of rain are landing on my face. But whereas I'm trying to shrink inside my clothes to get as far away as I can from the puffin-friendly conditions, Stephen looks relaxed and entirely comfortable. As a nature-worshipping druid, his attitude towards the weather is like that of a proud parent of a toxic two-year-old. He's utterly blind to all its sharp and ugly edges because he is so completely consumed with love for it. He doesn't feel discomfort when Mother Nature dribbles down the back of his neck or bites him on the fingers because, he feels, she's an extension of himself and he accepts her unconditionally. It's quite touching. And also quite annoying.

'Now,' he says, 'have a look at this tree and tell me what you feel. Don't try too hard and don't think. You should go with the first feelings that you have because your senses kick in before your thoughts.'

It's a mature tree, but thin and knuckly. Its branches seem dislocated at their joints and the trunk twists uncomfortably upwards as if it's trying to escape from some agony that's scratching and biting at it from the inside. This, I think, is not a happy tree.

'Very good,' he says. 'Trees are usually very comforting things to be around, but this one isn't. This one has seen a lot. You might also feel a hot, tingly sensation in your toes. People often associate cold with dark forces, but when you feel heat – that's when you should be careful.'

'Why?' I ask, looking around me at the kids on the swings and the Mondeos and Kas that are queuing up at the lights outside the park, grumbling and fuming like angry bulls. 'Are there dark forces around here?'

He looks at me as I shiver, my fists nesting deep in the corners of my coat pockets. Small drips of water are forming on his pale, wide face and matting the hair on his forehead.

'We are now standing on what was Herrington Hall,' he says, as a drip on his temple turns into a trickle. 'It was pulled down because nobody wanted to be in it.'

He points down to some ruined stone steps that appear out of the ground. They're low, crumbled and long redundant, having been conquered years ago by grass and dog turds.

'These steps used to lead from the kitchen to the gardens,' he says. 'Somebody was murdered on them. So, you were quite right about the tree. It saw that murder and it can remember everything. That's what you're picking up on now. It's easy to dismiss those feelings and just pass them off as something ridiculous, like, "Oh, it's just because it's shadowy round here" – the human brain loves to rationalise. But you should ignore those thoughts and learn to trust your feelings. The human senses are very, very strong.'

'So there's nothing actually bad here now?' I say.

'No,' he says, 'what you're picking up on is an echo. Mind you,' he says, looking over at a conspiratorial gang of trees that form a small wooded area at the eastern end of the park, 'I have been worrying of late, because all the rooks have disappeared.'

I follow the druid down a muddy track and we leave the old boundary of Herrington Hall and enter a quiet, hedge-lined road. As we walk towards an old church spire that looms up towards the grisly sky, Stephen tells me that everybody has psychic abilities and he is mystified when people act as though they're clever for using them. I like Stephen. He has bear-like proportions, and radiates a soothing, tranquil confidence, so that even though he has holes in his anorak, plastic bags in his pockets and speaks using only one side of his mouth, you tend automatically to take what he says seriously.

We cross the road. My big toes start to numb over. As I listen to Stephen, I think back to Michelham and the room with the mysterious opened drawer. As soon as I walked into that space that night, everything went foreboding. According to Stephen, that was my senses sounding the alarm. And if the rods are to be believed, it certainly looks as though we did find something. But was it the spirit of Sir Thomas Sackville?

If so, it strikes me, I might have found a terrifying truth about the ultimate cruelty of death. Perhaps the afterlife isn't the paradise that the priests believe in *or* the perfect sleep of the rationalists. Perhaps, when we die, we stay here, and death is slow and earthbound and endless. Perhaps souls can become stuck and lurk around their rooms for centuries. Or maybe there *is* such a thing as heaven and what I'd witnessed at Michelham was hell. Perhaps Lou's ghost lights and the presence at the priory were damned souls that have been condemned to a pointless eternity, chasing their tails amongst the living. And maybe they can communicate, very weakly, using tools like Paolo's dowsing rods or Kathy's gold necklace. This, I think, is a worrying development.

After a few minutes, we reach a gate in a shoulder-high wall that's the colour of old bones. Its four crumbling arms embrace a large, fallen cemetery. It has worn gravestones, dead trees, cracked crucifixes and sad stone angels covered in pale lichen scabs.

We pause, for a moment, in silence. I look at Stephen, who is smiling in his ragged blue anorak.

'Stephen,' I say, 'I don't really like graveyards.'

'Well, they shouldn't scare you,' he says. 'Graveyards should be neutral. By the time bodies get this far, their spirits are long departed. The only time graveyards are dangerous is when people have been summoning up.'

He points to a far corner, to a huddle of high, looming pines. A gang of rooks swoop, flutter and caw in the grey air above them.

'What do you feel when you look in there?'

I decide to hedge my bets. 'It's shadowy,' I say.

'That's right,' Stephen says. 'There is shadow there, and it's not all coming from the trees.'

'It's not?' I say.

'Trees can be corrupted,' the druid says. He turns and walks through the flaky metal gate into the cemetery, making his way briskly towards the corrupted trees. 'Remember what I said about summoning up?' he calls back to me. 'Well, this is a particularly spiritually dangerous graveyard to be in. Do say if you don't want to carry on.'

Standing in the centre of the copse, on a spongy matting of dead vegetation, Stephen tells me that he was walking down a nearby road one night when he saw a cloud hovering directly over these trees. Four dead-straight bolts of lightning shot out of the cloud and met just where we are standing.

'I don't know what was being summoned,' he says, 'but that was quite a powerful energy and I've never forgotten it. Lightning does not travel in straight lines unless it's called.'

For some reason, though, I feel fine. Perhaps my psychic antennae are exhausted. Then again, perhaps it's nothing paranormal. Maybe the presence of Stephen has reassured me. And maybe I usually feel foreboding in graveyards simply because I associate them with death and zombies.

'Right,' Stephen says, rubbing some rain out of his eyes with the back of his hand. 'There are some fields nearby. Do you want to have a go at some psychic protection?'

'Yes, please,' I say, and we troop off out of the graveyard, down a thin country lane, and clamber over a muddy stile.

'The first two we're going to try,' he says, as we arrive in the middle of an overgrown grass field, which is surrounded by low hills and bisected by a giant march of electricity pylons, 'are based on ancient Middle Eastern techniques that come from the Wisdom of Solomon. The first technique is called the Armour of the Soul. Close your eyes and picture your aura as bright and wide and burning all around you. Then, slowly, mentally bring it inwards, so it's only an inch away from you. Then turn it blue.'

I stand there, in the middle of the drizzly field, with my eyes closed, imagining myself glowing with a dense blue crust of light. Some time passes. A plane buzzes slowly overhead. Somewhere to my left, a sheep starts laughing at me. I half open one eye.

'Very good,' Stephen says, smiling proudly and nodding. 'Very good indeed. Next is the Shield of Tranquillity. This one has the benefit of being directional and is effective in any circumstance. If you're out on a Saturday night and see an awful mob down an alleyway – set it up. It's worked for me many a time.'

Following his instructions, I raise my hand and imagine a scorching beam of white light streaming out of it. I aim it at some rabbits.

'Excellent,' says the smiling druid as the rabbits run downhill. 'This last method is ancient Babylonian. It's called Circle Casting. Close your eyes and picture a sphere of white light around you that nothing can penetrate. Now, I'm going to go over there and walk towards you slowly. Keep your eyes closed, and when you feel me touching your aura, say, "Now."'

My socks, by now, are mostly completely drenched. My toes are fizzing or dead. There's a spreading patch of damp up my right trouser leg, where a sopping undone lace on my left shoe has been hitting it. Obediently, I picture a buzzing wall of pearly energy surrounding me. Some more time passes. Then, I secretly open one eye and sneak a look at Stephen to check what he's up to. I watch him for a second, inching forward through the long grass with his eyes scrunched shut, stroking the air with his hand as he goes. I close my eyes again.

'Now,' I say.

'Very good,' he replies. 'This protective shield is what stone circles maintain permanently.'

'Will I need to use these to protect me if I'm doing divination?' I ask.

'That all depends on the divination,' he says, walking towards me. 'The only dangerous tool is the Ouija board. That's an unfinished device. It opens the door to allow anything to come in.'

'What about pendulums or dowsing rods?' I say. 'I was told that they're dangerous because you're asking something to possess you.'

'Oh, no,' he says, smiling, basking his face in the miserable wet air. 'They're harmless. They're not possessing you, they're possessing the object.'

As we squelch our way out of the field, I tell Stephen about my mission.

'I'm trying to find the truth about ghosts,' I confide.

'Oh, that's all to do with iron,' he says casually.

'It is?' I say.

'Yes, it's often called the Stone Tape theory. We know that ferric oxide or chromium dioxide can record information. We use this process in everyday things like videos, cassettes, things like that. Now, iron is everywhere in nature. It holds information, if it's magnetised in the right way. So, very probably, ghosts are just repeats of information that is stored in natural iron. It's as if someone's rewound the tape and just played it again.'

Stephen explains that if we die a traumatic death by beheading, say, or pace up and down a corridor in a state of extreme distress because the Roundheads are coming, then this information is somehow stored in the natural iron around us. Then, elements of our situation can be recorded and, under certain circumstances, re-transmitted as headless ghosts or disembodied footsteps. Maybe, I think, people that we call 'psychic' simply have the ability to pick up on these natural recordings better than others. It's a compelling theory. And the best thing about it is that it promises a rational, unspiritual explanation for the existence of ghosts. It's absolutely God- and afterlife-free.

3

'Strange patterns'

I was expecting the British Library to be a huge, crannied and spired book-cathedral. It would be riddled with corridors and ante-rooms, all dotted about with green leather, inkwells and filled with still, church air. I'd work in hallowed silence in a long-forgotten ghost wing, the cuddle of light from my desk lamp the only illumination in the room.

But it's nothing like that. From the outside it looks like an immense warehouse on an industrial estate, built only for the eyes of fork-lift drivers and twilight hauliers. It's all sheer redbrick walls and no visible windows, and its only architectural feature is its brutish featurelessness. It's a joy-killing place of deadeningly functional aesthetics. If the British Library was a book, it would be the Yellow Pages.

And it's no better on the inside. It feels like a drab municipal sports centre, filled with students and their dreamy musk of stale alcohol, cheap food and sex. Worst of all, instead of there being a long-forgotten ghost wing, there's a computer screen into which I have to type to order my choice of book. It's like Argos.

I'm here for the purposes of research. Now that I'm fully equipped with four psychic protection techniques, I thought it was time I tooled myself up with some proper knowledge. For half a moment, I thought that Stephen the Druid might have nailed the whole ghost problem with his Stone Tape theory. But then I realised that this only really explains replay hauntings. It doesn't actually discount any of the things I witnessed in

Philadelphia. For that, I need the assistance of the massed ranks of the planet's ghost researchers. And I need their help quick. Because for me, on the day to day, things are starting to get weird. The more I explore this murky and ancient cellar of the human experience, the less everything else makes sense. The steely, rational scaffold poles that used to structure my ordinary life are slowly breaking down, and through all the holes that have appeared in my world, the ghosts are swooping in. I can feel their chilly slipstreams on my skin as I lapse into yet another daydream in the office. I sit at my desk and ruminate on death, hell and my earthly moral legacy, and I wonder if they can see you in the bath.

By the end of the morning, I've gone order crazy, and there's a toppling pile of ghost knowledge on my desk. The first thing I learn about is crisis apparitions. These are one of the most commonly experienced ghostly phenomena – and the saddest. They involve a person seeing the spectre of somebody at the moment of their death. Usually, these ghosts appear to people that they loved in life. It's a final goodbye, often accompanied by a soothing reassurance that death is not life's windy, desolate terminus. Diana Norman, wife of famous film critic Barry, has spoken about a 'smart, intelligent' friend who told her this story: 'I was at school at a convent. My favourite teacher was a nun called Sister Bridget. Then I got ill and had to spend some time in the convent infirmary. One day I looked up to see that Sister Bridget had come into my room. She smiled at me and I smiled back. Then she turned away and just walked through the wall. I began to scream at that, and the nuns came running in to see what was the matter. I told them that I had seen Sister Bridget and that she had disappeared. Then they told me that Sister Bridget had been taken ill and had died, a few minutes before I saw her.'

On the opposite end of the horror-scale are poltergeists. These are the terrifying bully ghosts that descend into people's lives and cause havoc in their houses. Poltergeist is German for 'noisy ghost' and one of their chief distinctions, as Lance told me in Michelham, is that they are people- rather than place-centred. Every case has a human 'epicentre', most often a teenager, usually a girl on the verge of negotiating puberty. In a book called *Deliverance*, an expert churchman called Canon Michael Perry explains:

> *The single most frequent cause of appeals for help are poltergeists. An attack often begins with small noises such as bangs, rattles, knockings, thumps or clicks ... after that, the effects rise to a climax and may include any of the following:*
>
> - *Rappings and knockings.*
> - *Objects may be seen to move of their own accord and in a bizarre way, defying the laws of motion and gravity, sailing in curved trajectories, or changing course at a sharp angle.*
> - *Usually, however, the movement itself is not observed, but only inferred from the results of it, so that the object is suddenly found in an unusual place only a second after it was seen elsewhere.*
> - *Objects appearing from 'nowhere'.*
> - *Glass, furniture, crockery being broken.*
> - *Doors being opened or closed; curtains billowing when there is no draught causing the movement.*
> - *Missiles directed at a person with great speed, but usually narrowly missing. Objects may refuse to move when watched, as if 'shy', but will immediately move when the observer is momentarily distracted.*
> - *Water dripping or pools of water appearing.*
> - *Rarely, spontaneous combustion.*
> - *Cold spots and, rarely, smells.*
> - *Sounds of music and jangling of bells.*
> - *Voices, or baby and child-like sounds such as sucking, smacking of lips, occasional crows and chuckles.*
> - *Very commonly, interference with electrical apparatus – lights switched on and off, domestic apparatus set in operation, record players and tape recorders made to fluctuate in speed, electric clocks working in reverse, telephones ringing for no reason, bells going haywire.*

Reports of noisy ghosts go back to Roman times and incidents are recorded in the medieval histories of Germany, China and Wales. One of the earliest fully documented poltergeist cases was investigated in 1877 by a founder of the Society for Psychical Research called Sir William Barrett. It occurred in the Irish home of a widower and his five children. Barrett discovered that if he asked the ghost to knock a certain number of times, it would do it. He then realised, to his astonishment, that it would even knock the right number of times if he merely *thought* the instruction.

In a meticulously researched book called *In Search of Ghosts* by Ian Wilson, I find a spectacular report from Long Island in America in 1958, in which flying objects and unexplained noises caused mayhem in a family home. All the scary fuss was witnessed by researchers and the local police, led by a Detective Joseph Tozzi, who went to incredible lengths to uncover the source of the disturbance. With the help of a small army of specialists, he tested for high-frequency radio waves, abnormal electrical activity, structural problems, water issues and local flight paths. He found nothing. The paranormal researchers who were documenting the nasty goings-on noted that only certain objects (including a record player, a night table and almost every bottle in the house) would be affected. This, coupled with the fact that the same things would be repeatedly thrown, signified that there was a process of selection going on. And that is the frightening part. Because it means that whatever was stomping invisibly around the house had intelligence.

Next I read about a haunting in a Cardiff car workshop in the early nineties, which appears to prove a link between poltergeists and apparitions. For over two years, many observers, including Professor David Fontana who is a fellow of the British Psychological Society and chairman of the SPR Survival Research Committee, saw a huge volume of phenomena, including bombardments of flying stones that would come from one corner and would last for hours, and coins appearing on request (all of which dated from 1912). The family who owned the garage and the shop attached to it assumed the spirit was a young boy because of its childish behaviour. This view was strengthened after an incident that followed the disappearance of a large rubber ball and a teddy bear from their shop. After hearing the sound of a bouncing ball coming from inside the suspended ceiling, the owner, Paul, got his ladder out and went up to investigate. The missing items were there. Then, towards the end of the haunting, Paul saw a full apparition of a boy, who was about twelve years old, sitting in the corner where the stones were thrown from.

If you take this evidence in conjunction with the findings of Sir William Barrett and others, it means that poltergeists are invisible, sentient ghosts of dead humans that can breeze through the laws of physics and read the minds of the living, and whose simple aim is often simply to persecute a person with massive terror.

Still in America, I find the true tale behind *The Exorcist.* The epicentre of the real attack was actually a thirteen-year-old boy, who had been experimenting with a Ouija board. It was 1949 when the family of Roland Doe first started to hear scratching noises in the walls. After that, furniture started moving, his bed would shake in the night and his sheets would be torn off, sometimes with the boy still inside them. He would also fall into trances, drool rivers of phlegm and would be mysteriously injured with scratches and cuts all over his body. A Lutheran minister, the Reverend Schulze, decided he was possessed by demons and led several exorcisms which, eventually, stopped the phenomena. In contrast with the film, however, Schulze managed to survive the encounter.

Another ghostly celluloid legend I looked into was Spielberg's *Poltergeist.* In primary school a friend once gave me nightmares by telling me that all the actors in the film had been killed by a curse. You can understand why I was scared. Just as in the movie, evil's malevolent claw had come right out of the TV screen and pulled the doomed innocents back in with it – probably to hell. Except, I now discover, that it didn't. Only four of the cast died early. The girl, Heather O'Rourke, died of intestinal stenosis in

1988, aged twelve. The woman who played her older sister, Dominique Dunn, was murdered by her ex shortly after the film's release. And Will Sampson, who played a 'good spirit', and Julian Beck, who played a 'bad spirit', died of heart and lung trouble and cancer respectively. All of which, it seems to me, lie well within the floodlit ballpark of simple coincidence. In the end, the most alarming thing I could discover about *Poltergeist* is that it was a PG.

The case behind another famous ghost movie is far more murky. In December 1975, the Lutz family moved into a house on 112 Ocean Drive in Amityville, the place where a man called Ronnie DeFeo had murdered his entire family. That much appears to be true. But the Lutzes then told the world they had suffered twenty-eight days of the most bizarre haunting phenomena, which included green slime spewing down the stairs, a spectral marching band, and possession of the girl by an evil pig called Jodie. Their spin-off book and film of these 'experiences' made millions. This resulted in law-suits ping-ponging about the place, including, interestingly, an action by Ronnie DeFeo's lawyer who claimed that he had approached the Lutzes with a concept for a ghost-based money-making scam in the first place and that they had done it without him. The Lutzes then counter-sued. In one of the cases, Judge Jack Weinstein stated for the record 'the evidence shows fairly clearly that the Lutzes during this entire period were considering and acting with the thought of having a book published'. The family who moved into Amityville after the Lutzes found nothing paranormally wrong with the house. The only thing that disturbed their peace was the constant stream of ghost-tourists knocking on the window. So, they decided to sue the Lutzes …

I move on. And am disappointed to find that *Ghostbusters* apparently has very little basis in reality. *Rent-A-Ghost* has none.

The Exorcist, though, unwittingly resembles a genuine incident from the nineteenth century more closely than it does its actual source. I find a report by a German doctor called Justinus Kerner who describes how an eight-year-old girl 'suddenly was tossed convulsively hither and thither in the bed, and this lasted for more than seven weeks; after which suddenly a quite coarse man's voice spoke through the mouth of this child'. Another case, in 1889, told of eleven-year-old Dinah Dragg: 'A deep gruff voice, as of an old man … instantly replied in a language that cannot be repeated here.' A different case again was documented, of a priest trying to exorcise a young girl. When the priest asked her, 'As you know so many things, do you also know how to pray?', the girl replied, 'I shall shit down your neck.' Just like Kathy Ganiel.

Then I come across the chilling story of a British man called Andrew Green. Born in 1927, he is one of the country's best-known ghosthunters. His father had been a rehousing officer and in 1944, when Andrew was in his teens, he accompanied him on an inspection of an empty property. Nobody had lived in 16 Montpelier Road in Ealing for a decade and builders who'd been working in it had complained of the sounds of footsteps, things being moved about and doors slamming in places where doors had been removed. One group of workmen fled after three hours. So Andrew's father asked a friend at the Met to check out the history of the house. He reported back that there had been a murder there, of a baby that was thrown from the property's seventy-foot tower. And there was more. There had been twenty suicides at the house, all of which involved people hurling themselves off the same place.

Their curiosity ignited, Andrew and his father went back to Montpelier Road for a poke about. When Andrew reached the stairs that led up to the tower, he felt invisible hands pushing him up. On reaching the top, he heard a powerful voice in his head telling him the garden was only twelve inches below and that he wouldn't hurt himself if he jumped. Andrew was only slapped aware of the deadly height of the drop when his dad realised what was about to happen and yanked him back from the edge. He'd saved his life. From that moment, Andrew Green has been obsessed with ghosts.

Green's investigation into Montpelier Road never reached a satisfactory conclusion. But a chance meeting with an ex-maid, years later, did suggest a possible occult explanation. She told Green that she remembered a strange routine that took place every Friday. The butler would leave the tower holding two silver candlesticks with black candles in them, as well as a mat covered in 'strange patterns' that she had to clean. Was this evidence that satanic rituals were taking place in the tower? Had the owners, with the connivance of the butler, summoned something evil that had remained, lurking amongst the dust and corners of the abandoned house?

One of the most disturbing aspects of my time in Philadelphia was Lou's live recording of Electronic Voice Phenomena. So after a lunch of two sandwiches, one coffee and some uneasy rumination, I decide to leap haphazardly into their history. EVP were first discovered in 1936 and, even though huge amounts of research has since been carried out on the subject, in Spain, Italy, Germany, Portugal, France and America, still nobody knows where the voices come from.

It was twenty-three years after their discovery, in 1959, that EVP became an international phenomenon. A Swedish documentary-maker called Friedrich Jurgenson was taping some twittering birds in his garden for use in one of his films, and when he played his recording back he was annoyed to find the sound of a man talking about 'nocturnal birdsong' in Norwegian all over the top of the tweeting. So, he did it again. And again. And the voices kept coming. Eventually, Jurgenson realised that they didn't belong to any living human. And his curiosity turned to shock when the voices started talking directly to him, using his familial pet name and mentioning the names of several dead relatives.

Still, sceptics insist that EVP are just roaming radio waves that have become lost in the ether, or meteors bouncing around the earth's atmosphere. This is despite more recent research by an American called Raymond Bayless, who has recorded voices under extremely rigorous laboratory conditions, and the Galileo

Ferraris Electrotechnical Institute in Turin, who found there was a complete absence, in the voices, of the fundamental frequency that is emitted by human vocal chords. This finding has been backed up by a spaceman. Alexander MacRae worked for NASA on voice authentication programmes on their Skylab project and has confirmed this, as well as discovering several other key differences between EVP and earthly voices.

Even minor celebrities aren't immune to the phenomena. When comedienne Sandi Toksvig went on location to a haunted castle, to record an episode of the show *Excess Baggage* for Radio 4, she and her crew were staggered when they heard strange whispers on their tapes. And this wasn't the first time EVP have found their way onto the radio. In the mid-eighties, a consultant electrical acoustic engineer called Hans Otto König tried to make a machine that enabled a two-way conversation with the voices (this has been attempted before – in 1977, an electronics engineer called Bill O'Neil had some success with a device he called a 'Spiricom'). König was invited onto Radio Luxembourg (which was, at the time, one of Europe's premiere radio stations) to demonstrate his invention to a sceptical crew and the DJ, a man called Rainer Holbe.

Holbe and a team of technicians watched König set his kit up and, for extra anti-fraud safety, the technicians refused to let König anywhere near his machine during the broadcast – only they were allowed to touch it. Then, during the show, one of the Radio Luxembourg crew asked if there was anybody there, and everybody heard a voice saying, 'Otto König makes wireless with the dead.' It answered the next question by saying, simply, 'We hear your voice.' By this point, the DJ was physically shaking. He told the listeners, 'I swear by the life of my children nothing has been manipulated. There are no tricks. It is a voice and we do not know from where it comes.' Following the broadcast, Radio Luxembourg released a statement confirming that the show was fully supervised and that all the technicians and engineers present were convinced that the EVP were paranormal.

There have been more recent experiments with two-way EVP conversations. Dr Anabela Cardoso, one of Portugal's most senior female diplomats, began experimenting with the voices in 1997. Instead of trying to get them to show up against a background of the 'white noise' that you find on a cassette tape, she used a detuned radio. She's chatted with several people who were previously unknown to her, and also many family members. Recently, Professor David Fontana (who also investigated the Cardiff poltergeist) travelled to Lyon, where Dr Cardoso is stationed as consul general, to listen to her tapes. He confirmed that the voices were loud and clear, and that many of them give their names, comment on events in Dr Cardoso's studio and have given information that she didn't previously know – such as the full name of the maid she'd just hired. (An interesting side-note: Cardoso sometimes talks to the EVP about the nature of the afterlife. They've told her that survival is a natural law for all beings, and that suffering in this world is important for spiritual development.)

EVP can come in any language, and you can get men, women and children. Sometimes, they even sing. In the early seventies, an SPR researcher called D.J. Ellis conducted his own investigation into EVP. After two years, he concluded that they were 'highly subjective' and existed only in the imaginations of the desperate listeners. All this puzzles me. Because what I heard in Philadelphia wasn't my imagination. And it wasn't an actual talking voice, either. In fact, we couldn't make out what *any* of the voices were saying. Not really. All that was apparent was that they were voice-like and that they sounded evil and angry and scary beyond words. Many other famous EVP experiments, including Dr Cardoso's, have produced solid, unequivocal voices. The ones that were broadcast on Radio Luxembourg were said to have sounded just as clear as those of the DJ.

As technology has grown and more electronic devices are invented, the spirits of the dead have discovered new ways of saying hello to the living. In 1985 a startled German called Klaus Schrieber received contact from two of his dead wives on his television. Voices have also been picked up on PCs, radios, telephones and videos. Computer printers have even been known to spontaneously spew messages from the deceased. I look around me, at all the students in the library. I watch them reading, their internal voices all chattering silently to themselves as their eyes pick over the words. I imagine the air around me filled with more chattering tongues than there are in the room. I start to sweat slightly, and decide to stop imagining things.

In 1967 the ghost of Thomas Edison appeared to a German and announced that anyone could contact the dead simply by tuning their TVs to 740 megahertz. That didn't work. But, when he was alive, Mr Edison *had* been convinced that the afterlife could be contacted. In 1920 he wrote in the *Scientific American* that he was working on a machine that could do just that. Unfortunately, his death came before the invention's completion.

Edison, I learn, isn't the only person of legend to believe in the supernatural. Sir Arthur Conan Doyle, author of the Sherlock Holmes books, was a believer in Spiritualism, the ghost-based religion. During a séance in Wales, a grief-ravaged Doyle was convinced he heard the voice of his dead son, Kingsley, speaking to him from heaven. And then, two years later, during a séance with a pair of celebrated mediums called William and Eva Thompson, he embraced the fully materialised spirit of his mother. However, a couple of days after this beautiful reunion, the Thompsons were caught red-handed with a load of wigs, fluorescent make-up and a dead-mother costume. Incredibly, this didn't dissuade Conan Doyle, despite his wife grumbling to him that his new hobby was 'uncanny and dangerous'. And he remained convinced until his death – an event his son witnessed. 'I have never seen anyone take to anything more gamely in my life,' said Adrian Doyle, who went on to claim that he was convinced his father would return to see his family. 'Why of course!' he said. 'There is no question that my father will speak to us just as he did before he passed over.' They were so convinced of this that an enormous 'Sir Arthur reunion' night was arranged at the Royal Albert Hall. An empty chair was left for Doyle in the middle of the huge Victorian coliseum. Sadly, he didn't show up. As a postscript to all this, it's probably worth pointing out that, when he was alive, Doyle also believed in fairies.

A far less ludicrous but equally renowned paranormalist I read about is the Swiss psychiatrist Carl Jung. Growing up, Jung was surrounded by so many supernatural happenings that his mother had to keep a logbook of all the inexplicable events that would take place in the family home. Despite this, Jung saw a ghost only once. He had a long stay in a haunted house and experienced gradually worsening phenomena. On the first night, he heard water dripping. Then he smelled something strange. Then, towards the end of his stay, while trying to sleep, he was assailed with creaks, bangs, rustles and raps on the walls, both inside and outside the room. When Jung opened his eyes he saw a head on the pillow next to him. Half its face was missing and its sole eye was staring at him. He was so scared that he had to spend the rest of the night with the candles on.

On reflection, young Jung decided that the experience had been triggered by the smell, which had reminded his subconscious of an old patient of his. He claimed that the knockings were his heartbeat and the rustlings were miscellaneous room sounds that were amplified and morphed by the power of the hypnogogic (half asleep) state. He was stumped by the dripping water, though, because he was wide awake at the time and a thorough investigation of the area revealed neither a water source nor a mechanism for its drippage.

In 1919, Jung gave a lecture to the SPR, in which he claimed that there were three sources for belief in ghosts – dreams, apparitions and 'pathological disturbances of psychic life' (which, simply translated, means – I think – being a bit mad).

Abraham Lincoln believed in ghosts and so did Jesus. The King of the Jews spends a good deal of time in the Bible exorcising possessed people – who were presumably poltergeist epicentres. There's even a quote in the Book of Job (4:15): 'Then a spirit passed before my face; the hair of my flesh stood up.'

Next, I set out to find ghost stories from the most distant reaches of history and geography. And I soon discover that the whole world is haunted. In ancient Egypt, the high priest of Amun was visited nightly by the irritable spirit of the ex-chief treasurer to King Rehotep. Chronicles from Roman times tell of a struggling philosopher called Athenodorus who bought a house on the cheap, only to find that the reason for its bargain price was there was someone dead living in it. The ancient Greeks, too, suffered from lingering dark traces. The site of the Battle of Marathon was a no-go area for years as 10,000 men were slaughtered there and it was as if the sound of the massacre became soaked into the wind that blew around the blood-drenched acres. 'Every night,' wrote witness and chronicler Pausanias, 'you may hear horses neighing and men fighting.'

Over in China, a friend of Confucius called Chu His wrote in 5BC, 'If a man is killed before his life span is completed, his vital spirit is not yet exhausted and may survive as a ghost.' Tibetan Buddhist funeral rites include post-mortal ASBOs that prevent dead people from doing any haunting, and some South American tribes put thorns on the feet of corpses to prevent their souls from rising and committing any upright mischief.

In Central and South America, they dread the 'Cihuateteo'. Jamaicans have their 'Goopies', Mexican tribes 'Tlaciques', Haitians 'Bakas', the Bantus of Zambia the 'Mudzimu', the Zulus 'Inzingogo' and in Guyana the 'Sakarabru'. The ruins of Babylon, in the south of Iraq, are haunted by two spirits known as Utukku and Ekimmu. In Taiwan, in the Tianan Taoist Temple, an apparition of a man is seen to this day, standing beside a huge stuffed crocodile. Visitors to the Kali Temple of Dakshineswar in Calcutta regularly report seeing and feeling the presence of Ramakrishna, a kindly man who was head priest in 1856.

On the other side of the planet in São Paulo, Brazil, in 1999, the council had to call paranormal investigators in to their head office after employees experienced apparitions disappearing in the middle of hallways, disembodied whispering voices, footsteps and phones that would ring and ring only for nobody to be on the other end. In Guanajuato, Mexico, the Museo de las Momias is filled with corpses that have been dug up from the nearby Pateon cemetery. Relatives of the deceased that lie in the mountain-top plot have to pay a grave rent, and those who default have the bodies of their loved ones removed and displayed in the museum. In 1969, the ghost of a woman was seen acting frantically in the area of the gnarled exhibits on the baby shelf. Witnesses said it was as if she was searching for her child. More regularly, visitors and staff members often say they hear voices in the empty rooms, as if the corpses are whispering to each other.

Singapore is said to be the most haunted country in Asia, but Japan is also crammed with ghosts. On 23 June 1590, Hachioji Castle in Tokyo was attacked by the fearsome army of Toyotomi Hideyoshi. Hachioji's female inhabitants, desperate to avoid the inevitable rape and torture that would precede their soon-coming deaths, flung themselves en masse off the ramparts and onto the jagged rocks below. After this horrific event, the castle was deserted for 400 years, because nobody could bear to hear the sound of the screams and the thudding of bodies on the rocks that still echoed through the valley. The sound can be heard, I read, to this day.

Those wars, though, are ancient. Vojvode Putnika Boulevard in Sarajevo, which became known as Sniper Alley during the recent conflict in Bosnia, is said to be haunted with new ghosts from the nineties. Similarly, nearby in Bijeljina, the spirit of a man who was killed by a Serbian strike force is said to stalk the night streets. He's been identified by witnesses as a Muslim called Mehmed.

And Muslims, it seems, do suffer from the same nightmares as Christians. To my surprise, I discover that exorcism isn't, as I'd assumed, just a biblical tradition. I find a report in a newspaper from Kuala Lumpur about the long tradition of Muslim exorcists, who recite verses from the Koran before demanding that the demon leaves its writhing victim.

Finally, at the end of the day, I pick up a book called *This House is Haunted*, by Guy Lyon Playfair. It's a first-hand account of the poltergeist case in Enfield, north London in 1977. It documents things that happened to the author and his fellow investigator, Maurice Grosse. Maurice, I remember, is the man that Lance from the Ghost Club called 'our most distinguished living investigator of polts'. This is the third time I've stumbled across this troubling story. And, remembering what Lou Gentile told me about things that happen in threes, I open the book gingerly and decide to pay this case the strictest attention possible.

'Hello, Andrew Green.'

'Hello, Andrew. My name's Will Storr and I'm doing some research into ghosts. I wonder if you could spare the time to meet me and have a chat about your experiences.'

'Did you know that somebody is writing my biography?'

'No, I didn't. That sounds fascinating.'

'Yes.'

' ... Is there any chance that we could meet, then?'

'I don't see why not. Would you be so kind as to give me a call back in a couple of weeks? I'm quite busy with things at the moment.'

'Yes, of course. I'll call you back then.'

'Many thanks. Goodbye.'

4

'Come back, Rain-On-Face'

Rain-On-Face is holding a pink gemstone up to a car window and examining it closely. He's a Native American – a member of the Eagle Owl Clan – and is counselling a young woman who has come to him for advice. Rain-On-Face is in his forties and wears a light denim shirt with four buttons open, revealing a captivating collection of beads and feathers on leather strings around his neck. He's a large man, and his bat-black hair is greased back into a tight ponytail. Rain-On-Face is also, I cannot fail to notice, Caucasian. He is strikingly, overwhelmingly, manifestly a white man. He's got pudgy, rosy cheeks, a stumpy, pug nose and a Geordie accent so thick it could curdle lava. Rain-On-Face's young student is worried because she put her stones under her pillow one night and when she woke up, one of them had gone cloudy.

'Did you dream that night?' asks Rain-On-Face, holding the triangular rock between his thumb and forefinger.

'Yes,' says the girl. She sounds astonished. She can't believe he knew that. 'Yes, I did.'

'Was it a happy dream?'

'Yes,' she says, smiling, 'definitely.'

'Then you should keep hold of that. What's happened is, the stone has caught your dream and now it's got some of your medicine in it. Here,' he says, handing it back, 'you should wash that in a light solution of salt water, because it's got my medicine on it now.'

The girl takes it back and stares at it for a moment, before stroking it softly with her thumb and pushing it deep into her denim jacket pocket.

I lean forwards to speak to Rain-On-Face. 'What do you mean by "medicine"?' I ask him.

'Medicine is a Native American term,' he shouts back to me from the front passenger seat. 'It simply means power, energy, spirit. So they talk about good medicine and bad medicine, if it's a good spirit or a bad spirit.'

There is a small pause. He turns round and gives me an accusatory look.

'It's got nothing to do with drugs,' he says.

I'm here on the invitation of Stephen the Druid. We're all squashed into a people carrier, en route to an investigation with a local paranormal research society called Avalon Skies to which he's affiliated. We're on our way to see Rosemary and Paul Astley, who live in a house that generations of Tow Law village residents have regarded as thoroughly haunted. Although they haven't experienced anything themselves, a friend of the family did see something astonishing when they hired him to clean the carpets. A former forces man and ex-prison officer, he was alone in the house while cleaning and turned around to see a little girl standing at the bottom of the stairs, watching him. When Paul came home from work, he said the friend 'looked very shaken – it was like cold water had been poured down the back of his neck'.

Some months after the sighting, Paul spotted one of Avalon Skies' distinctive 'Who You Gonna Call?' fliers in a Newcastle café and got in touch. The brilliant thing about this group is that they're like a band of cartoon superheroes. There's Stephen the Druid, Debbie the Witch, Trevor the Monsterologist, Rain-On-Face the Native American and a Voodoo Priestess called Michelle who can't be here tonight because she has a prior baby-sitting commitment. All of them have different spiritual powers that give them unique insights into the paranormal situations that they put themselves in.

After we arrive at the Astleys' neat, warm terrace, and I've had a chat with Paul, I settle down with Trevor, one of Avalon Skies' co-founders. I like Trevor. He's in his thirties and stocky, warm and chipper – a fair-haired, out-of-work plastics moulder and self-trained monsterologist. Most of all, I like the fact that – despite his monster specialism, and the fact that, when we met, the very first thing he told me was that he'd recently been hospitalised after a 'psychic larvae' attack – Trevor considers himself a sceptic. He formed Avalon Skies five years ago with his girlfriend Debbie (the Witch) when they found they had irreconcilable paranormal differences with the leaders of the group they belonged to at the time.

'The ideas they were coming up with were absolutely crazy,' Trevor says.

'What do you mean?' I ask.

'Well,' he says. 'You don't join a paranormal research society because you want to be in a music group.'

'A music group?'

'They were trying to get members to play instruments. They were totally taking it away from research.'

'That sounds absolutely crazy,' I agree. 'Were they getting you to play them in haunted locations?'

'No,' he says, the incredulity widening his eyes. 'Just as a leisure activity.'

Trevor's interest in the supernatural stems from the sudden death of his best friend, Brent. They were seventeen when it happened and studying together in the sixth form. One night, during a drunken walk to a post-pub party, Brent collapsed.

'He just fell over in the street,' Trevor says, sitting and shrugging on the sofa next to me, 'and that was it. Lights out.'

Brent's heart had failed. In the months after his death, Trevor would often bunk off school to visit his old friend's parents. Brent's mother had a sister who was a keen Spiritualist and had been trying to persuade her, without success, to attend a service. Then, one night, Trevor had a dream.

'I was in the science block at school,' he says, 'and I turned to go upstairs and there was Brent, standing at the top of the stairs. I went up and I shook his hand and said, "Aaah, fucking hell, how you doing? Haven't seen you in a while." He said, "I'm all right, but I'm in limbo." And then I woke up.'

Trevor didn't think much more about the dream, and didn't mention it to anybody. Two days later, he visited Brent's parents.

'Brent's mam sits me down,' he says, 'and she goes, "I went on Sunday, to the Spiritualist church. I got a message." And I go, "Oh, what is it?" She says, "Brent's all right, but he's in limbo."'

Trevor didn't tell Brent's mother about his dream because he was worried about upsetting her more. She still, to this day, doesn't know.

I sit back on the sofa for a moment and absorb the facts of Trevor's tale. You could argue that the word 'limbo' might have been implanted into the subconscious of Trevor and Brent's mother while they were together at some point. Perhaps it was used in a TV programme that they both overheard, or in a newspaper headline. Then, coincidentally, their unconscious minds threw the word back up again in the context of their grieving. But that doesn't add up, because it was a Spiritualist medium who passed the message – and the word – on to Brent's mother. And she was a stranger, a completely independent third party, that neither of them had met.

'Fancy getting some chicken drumsticks, then?' Trevor says, and we walk into the busy, toasty kitchen, where Paul Astley is preparing himself a pre-vigil hotdog amongst the chattering members of Avalon Skies.

'Before I moved in here,' Paul tells me, leaning against his Aga, 'my attitude to ghosts and the like was "bollocks to it all". Even when the little girl was seen, I just dismissed it. But then me and Rose started watching *Most Haunted* on the telly. We'd watch that and go, "Ah, I cannot treat it as mumbo-jumbo now."'

Most Haunted is the most popular show on the satellite channel Living TV. It features an ex–*Blue Peter* presenter and a medium called Derek doing vigils much like tonight's. Lance from the Ghost Club spoke to me briefly about the show. He said it was 'aimed at people who are not necessarily very well educated'. I got the distinct impression that proper paranormal researchers consider it to be a bit silly.

'Don't you think that *Most Haunted* is a bit ... ' I ask Paul, looking for the right word. 'Funny?'

'I find it funny that they hardly seem to pick any orbs up,' he replies. 'I find that very odd.'

On the other side of the kitchen, Rain-On-Face is sitting with Trevor and Debbie at a table that holds a lustrous pile of freshly grilled hamburgers. I sit down with them.

'Here,' I say, 'what do you lot think about *Most Haunted*?'

Suddenly, a knotty tension materialises over the table, tangling the atmosphere between us. I wait for an answer. Silence. Then –

'Why are you taping this?' says Debbie, eventually. She's peering down at my tape recorder, suspiciously.

'So I can quote you accurately,' I say.

'You're not going to *Most Haunted*, then? You're not going to play this to them?'

'No!' I say. 'I promise.'

'Well,' says Rain-On-Face, coming to her rescue, 'my feeling is, there's a lot of stuff that ... ' He pauses and looks to Debbie for reassurance. 'Well, I'm just not sure where it's coming from, put it that way.'

'I want to know who writes Derek's autocues,' Trevor says, with a smile.

Debbie shoots Trevor a spiky, sideways look, as if to say: Nice one, Trevor. Yet again, with your big bloody mouth.

'What do you think of Derek?' I ask.

Debbie's speech then takes on a careful, diplomatic policeman plod. 'He's a very good showman,' she says. 'Let's just say ... he's got the pizzazz to go with it.'

'And the hair,' Rain-On-Face says.

'Yes.' Debbie nods. 'And the hair.'

There's another silence.

'I heard he gets twenty-five grand a show,' Trevor says.

'Put it this way,' Debbie says firmly, 'the people on that show ... well, they wouldn't get into our group.'

'Why not?' I ask.

She thinks for a moment, with the face of a cautious defendant.

'They're very ... flighty,' she says.

'Well, they would get in our group,' Trevor says, 'but they'd get retrained.'

'Do you think they fake stuff?' I ask.

Suddenly, the tension swells so thick that it threatens to push all the oxygen out of the room.

'That's not for us to say,' says Rain-On-Face. He gives me a look. 'But I've got my opinions.'

'They had an orb that was clearly a moth,' Trevor says. 'They said it was an orb but it had wings and it was fluttering.'

Debbie gives Trevor that look again and says sharply, 'It's not for us to say', bringing the subject to a definite close.

I have been assigned to Trevor's team, and we're scheduled to investigate the cellar. Just before we descend, Stephen the Druid walks into the kitchen with several different cameras around his neck. These are back-ups, he tells me, in case he comes across a very common phenomenon of a haunted location – instantaneous power drain.

'You need at least one camera,' he says, 'that is based on purely mechanical technology. I've seen brand-new batteries drain in an instant on many an occasion.'

Down in the cellar, we are watching and listening, quiet and alert. The night outside and above us has been swamped by a swift, dense and mixing fog. It hangs above the ground in great surly bales, saturated with glowing orange by the streetlights. It feels ominous and animate, like it's pushing in on the house and, shut down in this old brick basement, it feels as if we're hiding from it. To my immediate right is a blocked-up arch passageway that, according to Avalon Skies, has been the centre of some previous activity. Nobody in Tow Law seems to know where the passage leads, why it was built or why it's been sealed with rocks and concrete.

I settle down quietly in the dark and breathe in the silence. I peer out into the darkness. Watching. Waiting. Listening. Suddenly, Trevor lets out a thunderous, trembling fart. The two teenagers who make up the rest of our team start sniggering tightly. They're holding huge guffaws back, up inside their noses.

'Whoops!' Trevor says. 'Hang on,' he adds, 'I've got another one brewing.'

Another vast ruction explodes from Trevor's trousers. His bad medicine fills the small room.

'Name that tune,' he says, and the teenagers collapse, releasing their rolling laughter.

'Shit,' he says. 'I think I might have broken vigil conditions.'

This goes on for twenty-five minutes.

Later, during the team debrief, I notice that Rain-On-Face isn't with us. I sneak out of the kitchen and creep up the stairs. Eventually, I find him in a back bedroom with Stephen. It's dark. The room is illuminated only by the orange fog that's pushing up against the curtainless window. I can just make out a pretty bedside lamp on a chest of drawers, a framed picture of cartoon pigs on the wall and Rain-On-Face lying flat on his back on the floor. The druid looks up, sharply. I get the feeling that I'm not welcome. I crouch down in the corner anyway, and guiltily watch the action.

Rain-On-Face has 'gone under'. That is, he's deliberately got himself possessed by the ghost that's haunting this house in an attempt to discover biographical information about it.

'How did you die?' Stephen says.

'It's the pains in the ... ' Rain-On-Face says in a faltering, almost female voice. 'I was in bed all the time.' He speaks so quietly you can almost hear the fog above his voice.

'Do you come to the house often in spirit?'

'No.'

'Just sometimes?'

Rain-On-Face stays silent.

'Can you connect with any of the other spirits that are in the house at the moment?' asks Stephen.

'Lizzie.'

'Did you like Lizzie?'

'Mmm, walking around with her ... fancy tits.'

There's a silence.

'Fancy tits?'

'Thinks she's everything.'

The question and answer session creeps on in the eerie dark for another ten minutes. As it does, Rain-On-Face's answers become more cracked, distant and whispered.

'Have you had enough, Mary?' Stephen eventually asks. 'Are you tired?'

'I don't like it.'

'You don't like it? You don't like talking to me? Do you find it strange?'

'Aye.'

'I'm going to leave you in peace now.'

'Oh, good. He he he he. Ahh.'

'What was that?'

'Ye heard.'

'Rain-On-Face?' says Stephen. 'Rain-On-Face?'

But Rain-On-Face isn't stirring. Slowly, a panic seems to build in Stephen.

'I can't wake him,' he says.

He leans over and starts stroking his hair.

'Come back, Rain-On-Face, you've wandered a long way off the path. We shouldn't have done that,' he mutters to himself, 'he was gone for too long.'

Then he says, a little louder: 'Come back, Rain-On-Face. It's time to return to us. Step back, Mary, and let Rain-On-Face come forward. Step back, Mary. Step back.'

Eventually, Rain-On-Face speaks. 'Tell her to leave,' he says, as if straining to break out of a profound sleep.

'Mary, you've got to leave, now,' Stephen says. 'Mary, it's time to let go. Take my energy to return. Come back, Rain-On-Face back.'

When Rain-On-Face stirs, it's as though he's loosely grasping at the details of a rapidly lifting dream.

'There was this woman,' he says, rubbing his eyes, 'she didn't want to let go.' He props himself up on his elbow. 'Where is everybody?'

Two hours later, I'm sitting in Trevor and Debbie's front room drinking tea. The milky dawn has filled the house with a dank, grey light. I'm exhausted. Behind me, a green parrot in a wire cage is having a noisy episode. The walls of the living room are covered in posters of white wolves howling at the moon. On the shelves there are heavy hard-backed books about magic and stone statues of dragons.

Trevor wants to show me some of the evidence Avalon Skies has gathered during their vigils at Newcastle Keep, the old fortress in the centre of the city. To be honest, I'm just being polite. At the moment, I really want to be by myself. I find it difficult, spending this much time with strangers. I feel like a bride who has to keep a piano-grin hoisted up for the whole day.

Trevor slides a homemade DVD into his machine and presses play. The static on the TV blinks off to reveal a green infra-red image of a stone staircase. The member who took the film thought that a wobbling sprite of light in the bottom left of the screen was interesting, but Trevor, being the sceptic that he is, quickly dismissed this as the unusual refraction of a nearby candle. It was only when he was watching it back, however, that he saw what I can see now: thin, beautiful wisps of light swishing down the stairs. They're delicate and quick, shimmering and weightless, and they vanish as quickly as they appear.

Then, a fat globe of light shoots down the stairs and disappears.

'Fuck!' says someone on the screen. 'Did you see that?'

'Fuck!' I say. 'Did you see that?'

'Aye,' says Trevor, 'it's neat, isn't it?'

It is startling. I ask him to send me a copy of the video and more footage he has of Debbie in a state of violent possession.

'Every time we go to the Keep we get something different,' he says. 'Do you want to come with us next time?'

'Urrrmm,' I say, transfixed by the footage that's replaying over and over again on the TV. 'Yeah. Yeah, all right.'

'Hello?'

'Hello. Is Andrew Green there, please?'

' ... Who is this, please?'

'My name's Will Storr.'

'I'm afraid Andrew passed away a couple of days ago.'

'Oh, Christ.'

' ... '

'I am so sorry.'

'That's OK.'

'Goodbye. Sorry.'

'Goodbye.'

5

'Distrust the mystic'

What are they, anyway? These strange machines, these units? Since the start of my journey, I've found myself watching the world in a different, more curious way. If you constantly look at everything through question-mark specs, things very quickly begin to get weird. And the weirdest thing of all is the people.

Honestly, if you view humans with enough distance, it really does get bizarre. I was having a meeting in the office this morning and, as I watched my workmates in action, I slowly became repulsed. In the end, I had to fight the urge not to run out of the room, screaming. Because gradually, my colleagues turned from ordinary, familiar people – Paul, James, Doug, Emma – into these horrific skin-covered engines, all interacting with each other, arms shifting, fingers slithering, eyes scanning up and down and left and right. They were machines, relentlessly gathering information through the senses and giving it out again, through this wet pink voice-hole that's rimmed with gory off-white food-blades. I sat there, spellbound and revolted, by these strange self-generating motors, all occupied with their own secret agendas and singular goals, chewing down their food-fuel and examining everything through their twitchy, blinking brain-cameras. What are these machines? Where did they come from? What do they want?

I've been pondering the nature of humans because I have become convinced, since I started thinking about ghosts, that *I* am not my body. I am the software and my body is the hardware.

If I'm to believe in ghosts then this *has* to be the case. Because if spirits are dead people, then it's this 'me' – this mind, this software – that escapes death. And it's just the body, the flesh-and-bone vehicle, that stops working. If ghosts exist, then we've all got a ghost inside us. It's the consciousness, the spirit, the soul or the mind. But is this right?

I decided to call the Royal Institute of Philosophy and tell them about the odd business that's been taking place in my head. By the time I'd finished, they'd agreed to despatch an Emergency Philosopher right away.

I wasn't expecting Dr James Garvey to look like a rugby player. I also wasn't expecting him to be clean-shaven or any younger than ninety-two. Most of all, though, I wasn't expecting my philosopher to be an American. To be honest, I thought, when we met outside the top-floor café of the Tate Modern in London, that I'd been short-changed. I wanted a wizened Greek hermit with a Crusoe-length beard and a staff. It quickly becomes apparent, though, that Dr James is excellent at his job – which is thinking. You can tell as much by his walk. He moves with a slow, graceful ease, as if he's thought thoroughly through the implications of each individual step. He has sad, dewy eyes, beautiful nostrils (I know it sounds strange, but really – they're like a dolphin's) and shoulders like a harbour wall. There is something hypnotic about Dr James. He's so rapt in his thoughts that a mist of serene detachment surrounds him. It's as if he has no worries at all – after all, what are worries but little, individually wrapped philosophical puzzles to solve?

We sit down with our coffees at window seats. From here we have a view over St Paul's, and the blue winter sky hangs, like a chilly ozone desert, over the spires, towers and lanes of old east London. I'm hoping Dr James will help me with my particular philosophical puzzle, which is this: if a ghost is the mind of a person that has survived death, then the mind must be a separate thing from the body. And my question is … is it?

'The view that the mind and the body are made of different

stuff,' he says, placing down his white china coffee cup gently, 'is known to philosophers as Cartesian Dualism. Descartes had the view that there are two distinct substances. There's a physical thing, which has properties, like it's located in space and it's movable. And there's the thinking stuff that doubts and wills and affirms and believes. One is physical and the other isn't. One is located in space and the other isn't located anywhere. It wasn't part of Descartes' main aim, in his meditations, to prove that mind and body are different, but he ends up doing it along the way.'

'Really?' I say, putting my cup down, too.

'Well, no,' says Dr James, 'he tried to. But unfortunately for ghost people, the arguments for Cartesian Dualism aren't very good.'

He then tells me what Descartes' main arguments are. And he's right. They aren't very good. The first one involves language.

Descartes was so impressed by our ability to speak to each other that he thought it was obvious something magical was running us. To him, it was impossible that a simple stimulus response machine could do something so complex, creative and spontaneous as to speak. So, we had to be divine, he thought. This, however, was before the invention of computers. But back then, Descartes, quite reasonably, thought that language proved we were special – unlike the stupid animals who couldn't speak and therefore *were*, effectively, just machines.

'For him,' Dr James says, 'animals didn't have souls. He thought that when you kicked them and they screamed, that it wasn't the sound of a soul in pain, but just air coming out.'

Just for a moment, I'm sure I can detect the first light wafts of a smile blowing in on the edges of Dr James's mouth.

'I don't remember their name,' he continues, 'but there was a group of monks who were vivisectionists, and they shared Descartes' beliefs. They would nail dogs down to the table and cut them open. "Oh, don't worry about the screams," they'd say, "it's just air coming out."'

'Oh dear', I say, 'that's awful.'

'Yes', says the philosopher. 'Yes it is.'

Then, Dr James tells me that the second argument is stronger than the first and it's called the 'argument from conceivability'.

'This said that whatever is conceivable is possible. And, for Descartes, it was conceivable that the mind and the body could come apart. And if it's conceivable that they could come apart, then they can't be the same thing.'

'Hang on,' I say. 'He said that anything that's conceivable is possible?'

'That's what he thought, yeah. So, you can't conceive of a three-sided square, can you? It's impossible to have one. The reason this is interesting is that, if you think that the mind and the body are the same thing, it should be inconceivable that the two could come apart. But it's not. That said, when you said "hang on", you were right. This is crap.'

'Was I?' I say. 'Is it?'

'Yeah. Because it's also conceivable that I've got tiny monkeys living in my ear. You see, just because it's conceivable, it doesn't mean it's possible. And also, what's conceivable depends on your background, your experience and all those kinds of things.'

So that's Descartes fucked, then. I take another sip of coffee and look out of the finger-smeared plate window again. A tatty stream of tourists trickle and eddy across the Millennium Bridge. And the sight of them gets me thinking. Could there really be a meaning to life? Are we, as the priests and the afterlife believers claim, all part of one giant, cosmic plan? From way up here, those people appear utterly pointless. What's the grand plan for that tubby old fella with the sunburned pate and blue jogging pants? Look at him, dropping his camera. Could he really be an essential cog in a complex, divine machine? I watch him bend down and pick up his camera. He rubs it against his anorak and gesticulates a grumble in the direction of his partner. Perhaps, I think, all this afterlife stuff is just a way of making us feel valuable and significant. From up here, it's hard to conceive that we're important enough to have souls. From up here, we look meaningless and silly and lost. Just a spreading rash on a rock near a sun.

Are there, I ask Dr James, any better philosophical arguments for independent souls?

'Yes,' he says. 'There's the claim that the soul doesn't have any parts. It's not physical, therefore, it can't be destroyed, so it can exist after the death of the body. And then there are a lot of contemporary arguments concerning zombies.'

'Zombies?' I ask.

'Yeah, can you imagine a world filled with zombies? That is, humans with no inner life?'

'Yes.'

'Well, we're not like that, are we?'

'No,' I say, glancing again at the dizzy little ticks of humanity on the riverbank beneath me. They may look a bit silly, but they're anything but zombies. They're jogging and snogging and bobbing about on the Thames in boats. There are hundreds of filled-up lives down there, all busy loving, worrying and bursting with thoughts and joy and tragedies. I imagine what it would be like if we *were* just machines. There'd be no art gallery here, for a start. And I wouldn't be sitting next to a philosopher, either, because there'd be no need for answers: we'd never have any questions that didn't involve food, drink, fighting and sex. The reason, then, that this functional zombie-world isn't a reality is because we *do* have an inner life, a soul.

'Exactly,' says Dr James, with a small nod, 'but there's a thousand problems with this argument as well.'

'But what about love?' I say. 'I can see the function of sex – that's to procreate. But why do we have to have love, as well? That's a uniquely human thing. That proves we're special, different, above the animals.'

'Love's for sex, isn't it?' says the philosopher. 'Maybe – and I hope your experience is different – lots of people seem to fall in love just long enough to have some excellent sex for a few years, raise a kid and then wander off. Love goes away, doesn't it?'

You can't argue with that. I've got several hundred CDs at home that sing sad, lonesome testament to the cruelly ephemeral nature of the heart. But there's another thing. What's the function of art? Art has no evolutionary purpose at all. You've only got to wander downstairs to work that one out. Art is the expression of emotion, from the artist to the consumer. Art's job, in other words, is to speak to the *soul.*

'I'd say that art is a kind of people-glue,' he says. 'It's a kind of thing that keeps people together.'

He takes another sip of coffee and shifts on his stool.

'There is something you could say about evolution and ghosts, though,' he says. 'It's hard to see how souls fit in to the evolutionary story, isn't it? Before humans evolved, were there souls just floating around waiting? And then, when we were ready, bang, they popped into our empty bodies? That would be weird. That doesn't fit. And there are other lines of philosophical thought that could apply to ghosts, too. A good one is from David Hume.'

Mr Hume, he tells me, occupied his thoughts with miracles, which he defined as anything that defies the laws of nature, like a man walking on a lake, or water turning to wine or, indeed, a person's mind surviving death and jeffing about the planet for eternity as a ghost. Hume saw it as a simple balance of scales. On the one hand you have the evidence for ghosts, which comes as testimony from mediums and witnesses. On the other hand, you've got all the evidence which suggests that people actually die when they die and that things can't move without something physical moving them.

'And evidence for that,' says Dr James, 'is everywhere you look.' He gazes around, at the customers and the food-trays and the plates of cake. 'Nope,' he says. 'Nothing's moving by itself. There's no dead people here. So the evidence for not believing in ghosts is, essentially, all the evidence you have for believing in the laws of nature, which are backed up constantly every time you look around. The evidence is overwhelmingly against it. It seems staggeringly unlikely that ghosts exist. So it seems like your default position ought to be: distrust the mystic.'

Hume's argument, that we shouldn't believe in the supernatural because clever people say that it can't be true, seems to me to have one obvious elephantine problem. It assumes that all these 'laws' of science, which are laws that humans have written, are absolutely error-free and finished. They just simply *cannot* be wrong. But 'experts' change their minds all the time. Priests, for example, claim to be experts in interpreting the rules of existence that God has laid out. A few centuries ago, they thought the Crusades were a great idea. They changed their minds about that. And still, Catholics believe that homosexuality and contraception is a sin. How long will it be before those tricksy opinions are rethought? It's inevitable, surely. And scientists used to think the world was flat. They changed their minds about that. These days, every single time they send a space probe up, they're forced to have a rethink. Really, every generation thinks they know everything, but everybody knows that they don't.

Dr James smiles at me, patiently. 'Sure, it would be wrong to say that we know everything,' he says. 'But your counter-argument, if that's what it is, is saying that for all I know, the current theories might be false, and that means that ghosts could exist. That's a kind of argument from ignorance. It's true to say that we don't know everything – monkeys might fly out of my ass, who knows? It might happen, but that's not a good reason for thinking that it *is* going to happen. But, I'm being facetious. What you said is true. We have been wrong in the past. We've been wrong a lot. But it's not entirely rational to conclude from that that ghosts exist.'

I find myself picturing Dr James with monkeys all over him. Tiny ones in his ears, flying ones down his trousers, a furious king silverback pounding on his head. Even if this man was ambushed and brought down by a wild swarm of deadly apes, he'd still be able to beat me in any argument.

'One of the things that killed Cartesian Dualism,' he continues, 'is the problem of interaction. You've said that a believer in ghosts would have to say something like, "Ghosts aren't bodies.

They leave bodies. A soul is something else. It survives the destruction of a body." And if a ghost and a body are different stuff, then the question immediately arises, how does the one affect the other? The kinds of causal interactions that we understand are things like billiard balls smacking into one another. So, how can a thing that doesn't exist in space, like a ghost, have an effect on something that does?'

'So, what you're basically saying is,' I say, picking up my coffee cup and making it float around in front of his face, 'if a ghost is made up of pure soul, it shouldn't be able to do this?'

'If the ghost isn't physical, no.'

I put the cup down again and have another think. I remember the philosophical point that Lance from the Ghost Club made. It was something about nature being an economical system. And that there's nothing economical about making us accrue all this wisdom, only for us to lose it all by dying, just when we've got life sussed. Lance is right. That just wouldn't make sense. Would it?

'Maybe,' says Dr James. 'But there are lots and lots and lots of things in the universe that end. There's a thing called the "argument by design", which is the view that everything in nature seems to be organised to achieve a certain purpose. Imagine you were walking along a beach and you found a watch. You would look at its parts and you'd see the whole thing has been designed to achieve a certain purpose. Somebody must have put it there. And you can look at the universe in the same way and say it's designed, that there must be a universe-maker, just like there is a clock-maker. You can see patterns. Maybe this Ghost Club person is seeing all this and thinking that it's all been done for a reason, that you couldn't accumulate all this information and then just die. Well, he's wrong – you can look at the earth and see all sorts of things that look badly, badly designed. You breathe through the same thing you drink through. How stupid is that? You choke because of that. Whole species go extinct. There are ice ages that wipe stuff out. There's disease.

People go blind. There's lots of unpleasantness in the world. Everything tends towards disorder. In fact, it's even one of the laws of nature. It's entropy – stuff falls apart. Something like a human being is a little momentary flicker, or a wisp of organisation in the universe, not part of a pattern. If you look at other human lives, like little babies writhing in chemotherapy, you see something that ain't designed at all. At least, not designed well. You know,' he says, looking at me with his thoughtful eyes, 'maybe all the information that people accumulate over a lifetime ... maybe that's it. Perhaps that's what ought to be celebrated. That bit there that we got. Hoping that there's more, looking for ghosts or religion or whatever, is wasting that little good bit that you've got.'

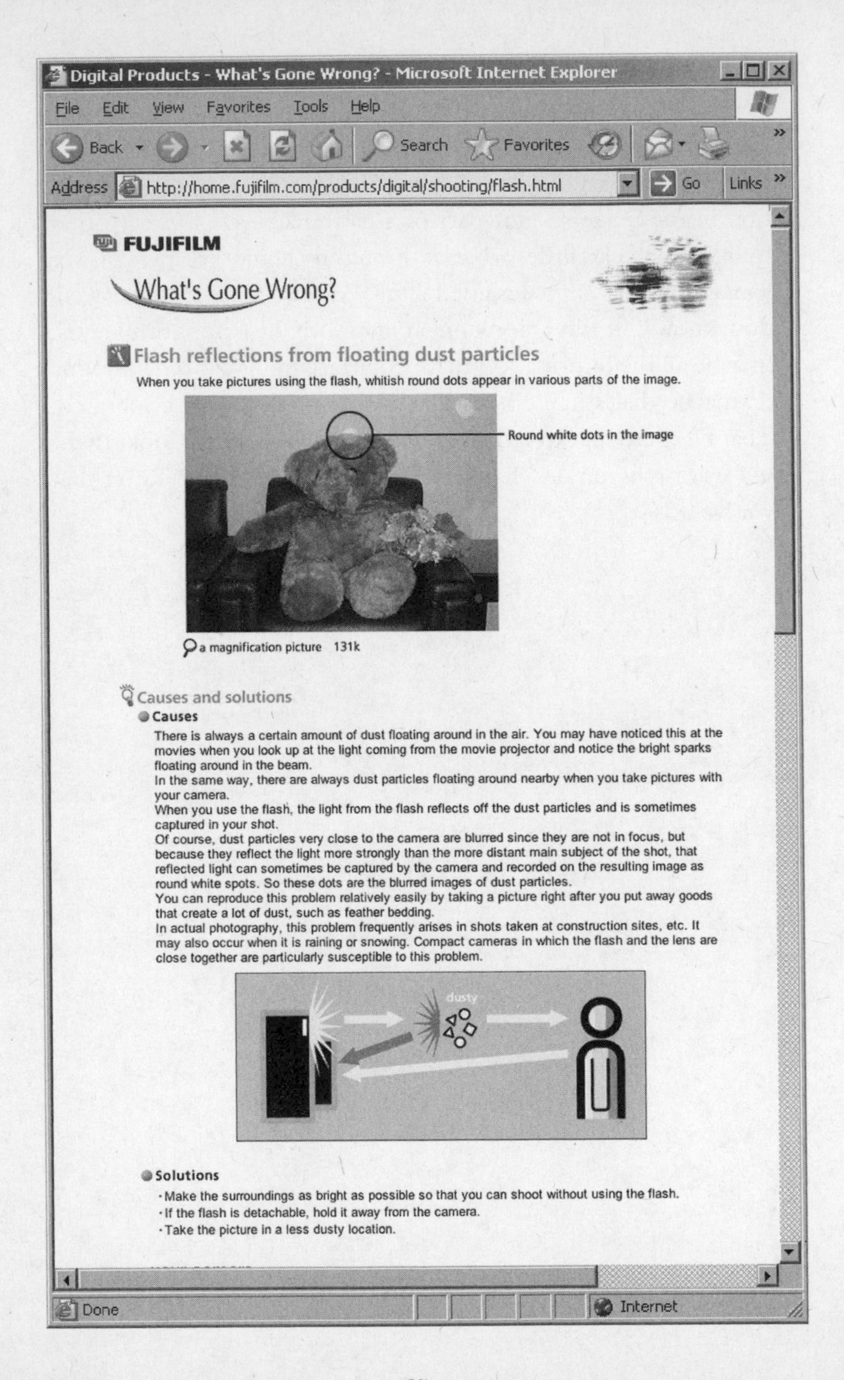

Digital Products - What's Gone Wrong? - Microsoft Internet Explorer

File Edit View Favorites Tools Help

Back Search Favorites

Address http://home.fujifilm.com/products/digital/shooting/flash.html Go Links

FUJIFILM

What's Gone Wrong?

Flash reflections from floating dust particles

When you take pictures using the flash, whitish round dots appear in various parts of the image.

Round white dots in the image

a magnification picture 131k

Causes and solutions

Causes

There is always a certain amount of dust floating around in the air. You may have noticed this at the movies when you look up at the light coming from the movie projector and notice the bright sparks floating around in the beam.
In the same way, there are always dust particles floating around nearby when you take pictures with your camera.
When you use the flash, the light from the flash reflects off the dust particles and is sometimes captured in your shot.
Of course, dust particles very close to the camera are blurred since they are not in focus, but because they reflect the light more strongly than the more distant main subject of the shot, that reflected light can sometimes be captured by the camera and recorded on the resulting image as round white spots. So these dots are the blurred images of dust particles.
You can reproduce this problem relatively easily by taking a picture right after you put away goods that create a lot of dust, such as feather bedding.
In actual photography, this problem frequently arises in shots taken at construction sites, etc. It may also occur when it is raining or snowing. Compact cameras in which the flash and the lens are close together are particularly susceptible to this problem.

dusty

Solutions

- Make the surroundings as bright as possible so that you can shoot without using the flash.
- If the flash is detachable, hold it away from the camera.
- Take the picture in a less dusty location.

Done Internet

6

'Making things fit'

For generations the parishioners of St Nicholas's have known about the sorrowful spectral forms that move in their church's graveyard. The ghosts have been here since late in the sixteenth century, when a madness passed through their village like the shadow of a vast cloud. And ever since the clamour and terror of those witch trials and executions, the villagers have known that this cemetery is stained. They say that if you come in the night, as we have, you can still hear traces of the things that happened then, in between the crow calls and the gusts that blow and whistle around the steeple and the stones. And it's then that you can sometimes see the victims, still mourning at the site of their ordeal.

The sun has long since set and the only thing moving inside the churchyard's boundary is the sandpaper winter wind that rises off the wheat fields and runs between the gravestones. There's nothing here but the gusts and ... oh yes, the TV crew. And the lorries. And the chattering fans. And the luxury caravanette. And the most famous medium in Britain, getting angry with himself under a brilliant white arc of light that a film crew are beaming down on him.

As the shrill breeze runs through his accurately groomed hair, and multiple diamonds sparkle on his earlobes and fingers, I watch Derek Acorah trying to pre-record a link for his television show. After his fourth fluffing, he gives himself a pantomime slap around the face and says, with a whinny, 'Oh, I can't act!'

The crew who are filming Derek and his co-presenter, Yvette Fielding, do not offer any comment. They just remain in position and wait for Yvette to begin, yet again.

'Good evening,' she says into the camera, 'and welcome to *Most Haunted Live*. This is the first night out of three and we're on the hunt for the ghost of the East Anglian witch trials. We're here with the *Most Haunted Live* crew and, of course, we've got Derek Acorah. Derek, what are you expecting to happen here?'

'I feel over the next three nights,' he says, in his rich, scouse-with-added-sugar voice, 'you know, it's going to be very eerie.'

'Ooooh, no,' says Yvette.

'Be prepared to be *very scared*,' Derek says, with his forefinger extended and slamming through the air to give the last two words maximum dramatic impact.

'Ooooh, no,' Yvette says again. 'I don't like the sound of that at all. It's going to be amazing.'

I have broken my promise to Debbie. I have come on to the set of *Most Haunted*. Although I am not going to play anybody the tape of our conversation, I have managed to talk them into letting me take part in this evening's programme – which is a special live edition. On a normal show, Yvette, Derek and their team would simply spend the night in a haunted location and try to film whatever phenomenon they can with their infra-red cameras, while Derek sniffs out bonus information with his clairvoyant powers. Tonight, though, there is a presenter, various experts and an audience sitting in a studio a few miles away from us. They'll be watching and discussing our experiences as we travel around Essex on the hunt for happenings. My role in the proceedings is to be a 'verifier'. I have to confirm to the audience that any eerie occurrences that Yvette and Derek report did actually happen.

I was interested in how extraordinarily guarded the members of Avalon Skies were when talking about the show. They were acting as if the *Most Haunted* secret police could have stormed in at any moment and dragged them all off for heresy. I was also struck by how impressed Paul Astley was with the programme. I thought it remarkable that *Most Haunted* turned him and his wife into believers, and not the ex-squaddie friend who saw a dead girl standing at the bottom of their own stairs.

As far as the state of my own belief goes, Dr James has ripped my brain in half, like a Tuscan farmer splits a fresh lump of bread. On the one hand, my trip to the British Library has affected me deeply. I was astonished at the quantity of evidence and its cross-time, cross-cultural nature. Most of all, though, I was amazed by the consistency of the claims. The fact that poltergeist reports, for example, have remained the same across the world for hundreds of years makes them all the harder to be discounted. And what's more, when it does seem likely that a 'haunting' has been invented, as at Amityville, it's simple to spot. Even apart from the trail of money, there's a string of clues throbbing right through that testimony. A demonic pig called Jodie? Rivers of green slime? A spectral marching band? The implausibility of reports like this only serves to bolster the plausibility of all the others.

That said, I've been equally struck with Dr James's argument about evolutionary theory not fitting with a belief in ghosts and souls. When humans were just hairy fire-worshipping rapists living in caves, were we an elevated species then? Are there Cro-Magnon men in heaven? Are there cavemen ghosts? What about when we were just slime, floating about in the primordial ooze, bumping into each other and absorbing swamp-shit through our see-thru skin? Where were our souls at this point? Were they, like Dr James said, just hanging around some astral lobby, waiting for the moment when nature was ready for them? And when the timer on that evolutionary oven pinged, when we'd been fully baked to human, did somebody release a big net and did the souls float down and find an empty, man-shaped vessel? If so, who pulled the switch? Who made the oven? Who put our unevolved dough in it and worked out the timings?

When I think about this – and the answer that so obviously presents itself – I get that vision, again, of my mother at the dinner table. *Oh, you will be.*

And something else has happened. The other morning, I got a package from Trevor with the DVDs that he promised me. The footage of Debbie under possession is fascinating. She appears to be in the grip of a man called George Turnbull, who, they discovered after much research, was a smuggler with a history in the pub where the possession happened. Debbie certainly doesn't look as though she's acting. The footage, of her shouting 'I will not be accused' in a baritone rage, is violent and visceral. And, unexpectedly, the few minutes following her 'possession' is just as affecting as what came before – Debbie seems genuinely confused and drained to the point of collapse.

And then there's the film of Newcastle Keep. After coming across Fuji's 'What's Gone Wrong?' webpage, I am now comfortably sure that the orbs that digital cameras pick up are not balls of wandering ghost energy. It's dust. However, the infra-red, moving footage of light anomalies that Trevor and Lou have caught is harder to explain. Firstly, because of the way it behaves. At Deborah Carven's house we saw just one globule, in all the hours that we watched, and it followed her up the stairs. And in Kathy Ganiel's house, there were so many of them behaving in such a bizarre fashion, that I cannot persuade myself to believe that they were dust or insects. But most of all, on Trevor's film, somebody on camera sees it float down the stairs. He shouts, 'Fuckin' hell! Did you see that?' Firstly, people don't usually react like that when they see dust. And secondly, for the Fuji theory to work, the dust has to be right up close to the lens. But the member of Avalon Skies saw it where the viewer sees it – on the stairs. So this isn't a trick of perspective.

But, anyway, right now, I have duties to perform for the nation's armchair ghosthunters. Much of *Most Haunted*'s fun revolves around ex-footballer Derek, who has a 'spirit guide' called Sam. Sam lives inside Derek's head and tells him things about the ghosts and historical events that have taken place in the area. Sam, it has to be said, has mixed results. Much play is made of the fact that all the locations are kept secret from Derek in advance of filming. This way, they say, we know that everything he comes out with is of supernatural origin. Of course, critics believe that Derek *is* briefed. Some speculate that Acorah types the location into a search engine before he leaves the house and memorises any names, dates and grisly tales that pop up in his browser – something he roundly denies. On at least one occasion, though, Derek has been caught mispronouncing a spirit's name in modern English and not – as he might be expected to if he was actually in contact with it – in the olde way.

Just as Derek and Yvette are finally managing to get their lines right, I decide to have a wander about. There are three big production lorries parked with their lights on by the graveyard. As I'm peering into one of the doors, I feel an ominous, looming presence rear up behind me. It's Judy, the P.R. woman from Living TV, who has been assigned to chaperone me. On the way here, she made me promise to behave myself. I think my wandering is worrying her.

I follow Judy obediently back to the graveyard. This place is intricately linked with the story of Canewdon and Matthew Hopkins, the Witchfinder General. One of the most common sightings is the ghost of a spectral grey lady.

Tonight's show is being anchored from a makeshift studio in a nearby village hall. I'm standing on the exact spot that Judy has directed me to, watching the crew listen to its progress in their earpieces. Next to me is James, a journalist from a daily newspaper who is going to help me with the verifying. So far, Judy's doing a great job with us. On the way here, she had nothing but bowl-eyed praise for Yvette and Derek. And, after catching me trying to peek at the official filming schedule for the evening, she ticked me off politely whilst clutching it to her bosom. After that, I noticed, she took it to the bathroom with her when she needed to pee.

Back in the graveyard, our verification skills have been called upon already: something genuinely strange has happened. Just before Yvette and Derek go on air, the battery on the portable floodlight drains instantaneously – just as Stephen the Druid said.

'Shit!' says Yvette, hopping about the place, as the crew run around finding torches to shine on her face so it will show up on the telly. 'The light's gone off! Shit, shit, shit, the light's gone off!'

Then, a small lamp that was illuminating a gravestone falls over.

'Hi guys, hi,' she says into the camera, and to David in the studio, as the live broadcast begins. 'Hello, everybody in the audience. I think we should just explain something, quickly. We've only just been standing in the graveyard for a short amount of time. Already the battery power on the light has gone, there's a light over there in the corner that's gone over on its own. But there's no wind here whatsoever. It's not nice at all. At all. We've even got some journalists here as well that can verify. It's not windy, is it, guys?'

It isn't windy. But then, the light was standing on a flimsy tripod on bumpy, grassy ground and …

'Not at all,' I say, to Judy's visible relief.

Next, we all follow Yvette to the fallen light and watch her crouch down on the floor. She fiddles about with it a bit, in a breathless manner.

'Now, this is quite a heavy light and I don't even know how it goes back up, but it really has come down. It's just weird. Derek, what do you think?'

'Well, we all heard it with the naked ear and it was a real sharp sound,' says Derek.

'Look! You have to twist it!' Yvette says, holding it up towards the camera and twiddling a black knob on the side of one of the tripod's legs backwards and forwards.

'Yeah,' says Derek.

'That's really strange. You have to twist that. So do you think there's actually any activity here at the moment?' she says, looking up at the medium.

'Yes, and I feel there is activity here – and it's not just residual energy – which I'm excited about, and I hope things are gonna start to happen.'

'If anything comes out at us now I'm gonna absolutely wet myself,' Yvette says, before handing back to the studio.

Everybody falls silent again. Derek frowns, rubs the end of his nose a bit and looks away, far into the tungsten-speckled distance.

'Come on, ghosties!' sings Yvette.

There's a long silence.

A twig snaps.

'Did anyone hear that?' says James.

'I did,' I say.

Judy, it has to be said, appears highly suspicious of her verifiers. She unleashes a dramatic frown in our direction. It's as if she thinks we're making the snapping-twig story up – until Yvette pipes up.

'I did, too!' she says.

We're all looking in the direction that it came from – a bushy ditch on the edge of the graveyard. A strange black shape moves through the space. It isn't fuzzy or indistinct, it's sharp and clear and undeniably there. We stand silently for a second and watch it travel in stilted, jerky movements. Could this be the Grey Lady? Could the psychic remains of this dead witch really be moving through the shadowy shrubbery, shaken awake by the noise and bother of Britain's premier ghosthunting TV crew?

Derek clearly thinks so. 'There she is!' he shouts, and runs right at it with his arm in the air in the 'charge' position.

As we run after the medium – who's now disappearing into the depths of the cemetery – it occurs to me that, as Derek isn't supposed to have been told that we're looking for a Grey Lady, 'There she is!' is a slightly odd thing to have said.

By the time Judy, James and I catch him up, dodging the wires and ankles of the kit-laden film crew, the shadow has disappeared. But we can still hear movement.

'Sshh!' says Yvette.

We're all dead still and listening hard. There's silence, then –

'Bollocks!' shouts somebody from behind the trees.

St Nicholas's Church is at the top of a hill that has a thin windy road leading up to it. From one direction it looks totally isolated – an old stone steeple rising from the middle of a lonely countryside hamlet – but in reality it backs onto a large, teeming council estate.

More twigs snap. Laughter. It's rogue teenagers. They've infiltrated the set of *Most Haunted Live*.

'Boorrrlloooooorrrkkkkksss!' shouts one.

'It's all a load of bollocks!' sings another.

'Mary!' shouts another joker, in a squeaky high-pitched voice. 'Mary! Oh, Mary!'

Yvette is visibly furious. She rams her fists on her hips.

'Tony!' she shouts to the security man who's been hired to keep intruders off the set.

We watch Tony lunge into the trees.

'Fuck off!' we can hear him shouting.

'Bollocks!'

They're still going at it when we're back on air but, luckily for the *Most Haunted* crew, Yvette and Derek's microphones aren't powerful enough to pick up their voices.

'Here we are in the graveyard at Canewdon, St Nicholas's Church,' Yvette says, excitedly, 'and I have to say, even while we've been off air we have – all of us, I can safely say – have seen what can only be described as a shadow. Also, the mobile phone that Ross has as well, his battery has completely drained. So, we're having batteries going down on us as well. So, already, it just seems to be a really, really strange place. So Derek, things seem to be happening, and you actually sensed something over there.'

'Absolutely, most definitely a female figure,' he says, to my rising bafflement. 'To me, it was a female that was either bent over, or she was very, very, very, very short and she seemed to be at that angle. And you know what I get with it now? I keep on getting it ringing through, psychically, here to the front of my head, coming from here, and I get "Mary, Mary, Mary, Mary" and I get … is that right, Sam? "Mary Moon".'

'Mary Moon?' says Yvette.

'Hold on!' Derek says, his face sucked in with the concentration. 'Mary, Mary, Mary, Mary, it sounded like, yes, it is Mary, I get Mary and I get the name Moon. Now whether the Mary is to link with the moon, or whether it's the name Mary Moon, now I feel because I'm concentrated on that area. That's connected with that area, so this could be the person that's moving about here and I feel she would be quite elderly, not a young spirit person, quite elderly.'

'Why is she here?' says Yvette, mournfully. She's hugging herself tightly in her big black puffa, as the wind plays with her hair. 'Why do you think she's still haunting a graveyard?'

'Hold on,' he says, as Sam feeds him the requested info. 'Hold on. Now I'm getting Margaret. I feel as if I'm being strung with the psychic link with Margaret and after being hung, then two souls taken away and disfigured, and taken at the neck, taken the head off, and that is the one who walks around here,' he announces finally. 'It's Margaret Moon.'

Next, we are led to the side of the church where somebody has noticed that the infra-red triggered security light isn't working properly. As Tony hurtles about in the undergrowth, trying to keep control of the mudlarks that have gathered in gobby gaggles behind hedges and fences, Yvette and Derek discuss the mis-flashing light.

'When spirit activity is about,' Derek says, 'lights, batteries, anything electrical can be charged because what the spirit people are doing, in order to try and manifest in the atmosphere, is to take energy from the light. And so that's why we've got batteries going down.'

'That's very, very common, isn't it,' says Yvette, 'for people who do say they have haunted houses? It actually drains the electricity and they have very, very high bills.'

'Yeah,' says Derek, ruefully, 'and people don't know how much drainage they're getting until they get the bill.'

After we've handed back to the studio, the crew walk back to the outside broadcast vans. Tony, the security man, catches us up as we get there.

'Those little wankers,' he says to the producer. 'They were going, "What you gonna do? The police are twenty minutes away." There's not much I can say to that, is there? Typical Essex attitude.'

Ten minutes later, we've walked to the bottom of the hill to a crossroads where, legend says, a headless witch was buried with a stake through her heart. When we get there, there's nothing to see but a traffic island, a road sign and four empty tarmac roads that fade quickly into the night.

'Do they expect us to be here for five minutes?' says Yvette, looking around crossly at the featureless scene. 'It's crap.'

'It's not crap,' says someone from behind the camera, 'it's poignant.'

A few seconds later, we're back on air.

'This is a very poignant place,' begins Yvette, 'and a lot of things have been seen here and people have particularly witnessed one spirit. Can you pick up on anything at all, Derek?'

Everybody looks at Derek, expectantly. The crew, the viewers and us verifiers are all now wondering the same thing. Will we get the money shot? Will Acorah manage to sniff out the name of the main ghost that's supposedly been kept secret from him – that of Matthew Hopkins, the Witchfinder General?

'I feel as if in this area there's a lot of misery,' Derek begins, cautiously. 'Feelings of misery. Feelings of hopelessness. Feelings of minds, of individual people being grouped together and they seem to be fearing that they're going in a direction to be, I dunno, the only way I can describe is to be judged or something? They're fearing the arrival. They know, they're being taken.'

'By who?' says Yvette as they stroll up the hill, with the crew in front of them, walking backwards.

'Who are they being judged by, please?' Derek closes his eyes as if deep in reverent prayer. 'OK ... OK ... being judged by Matth ... no, no, not judged by him. Taken by him. Taken in that direction, by ... '

'Who? Who? Who?' says Yvette.

'These people, it's like they're being taken by Matthew? To be – it's like he's producing these people for them to be, in some manner, judged. And in that judgement it's like as if he supplies, and what I get with this Matthew, as well, with this supplying, it's like, to me, it's like, supplying an animal for the slaughter. It's like as if a human being's meat in his eyes, and with it I picture this, like, very strong lying, lying, making things fit, making things fit, um, and in nine times out of ten probably in situations with this, lies were told to build a picture, a false picture.'

'So this is Matthew telling lies?' says Yvette, trying to pull the full name out of him.

'Matthew telling lies,' Derek says. He looks baffled, as if his own words are mystifying him.

Five minutes later, we still haven't got a surname for Matthew Telling Lies.

Eventually, Yvette asks, 'Can you give me a surname for this man?'

Derek addresses Sam again. 'Can you give me a surname?'

The medium listens to himself intently as he paces up the cold country road in his long, expensively cut black raincoat.

'I feel he would have been quite happy to wear a hat and also a gown,' Derek says, outlining a big square above his head.

'You've done like that,' says Yvette, 'so I'm presuming … '

'It's a tall hat, yes.' Derek nods. 'And he would wear something covering like an over-gown um … who's that? Say it again?'

Derek walks along for a while longer in silence. His face is intense, determined.

'I'm just asking stuff from him,' he explains.

Suddenly, Derek stops. We all watch anxiously. Then, he closes his eyes and throws his hands up to his mouth. His psychic synapses are white hot and sizzling.

'It is!' he says. 'It is! Is that him? It *is* him!'

Everybody gazes at Derek and waits.

'OK,' he says. 'Sam's just confirming this.' The medium stares at the air in front of his face as he listens to his spirit guide. And then, he's got it and he says it. 'This Matthew is Matthew Hopkins!' he announces, slamming his finger down triumphantly. 'Matthew Hopkins!'

The crew nod at each other and smile as a satisfied happy-ending glow spreads through them.

After Derek declares the year of Hopkins' death as 1647, and comes up with some names for the witchfinder's 'cohorts', Yvette asks him if he can tell us where Matthew Hopkins is 'grounded'.

'I'll try. I'll try,' he says. 'OK, help me with this Sam, please.'

'All right,' says Yvette with her finger on her earpiece, 'join us after the break to find out where Derek thinks he could be grounded. OK, keep thinking, yeah?'

'It's a building, it's a building. It could be the building is white. It could be that the building's got something to do with maybe a name. White ... '

'Just to let you know,' the producer says, 'we're off air.'

With that, Derek snaps out of his trance, digs into his coat pocket for a packet of fags and lights one up. He smokes it sullenly, looking out across the silent fields. After a couple of minutes, the programme begins again from the studio. Yvette listens in with her earpiece and gives us a running commentary.

'They're saying you got the date right,' she says.

'You got the name wrong, though,' the producer jokes. 'It was Bob!'

They both snigger.

I decide to take this opportunity to sneak away and climb back onto the coach. I tip-toe over to Judy's seat and shove her schedule down my trousers before creeping into the toilet to fish it out. Cramped in the plastic capsule and bathed in slimy yellow light, I scan down the crumpled page. With a start, I notice that Derek has 'happenings' scheduled in advance: 'Yvette and Derek explain local folklore and have a happening'; 'Yvette and Derek at St Nicholas Church – further happenings'.

Conscious of the time, and the wrath of Judy, I shove the schedule back down my trousers and sneak out again.

'Can I just ask our journalists,' says Yvette just before the credits finally roll, 'how did you feel that tonight went?'

'I think it went really well,' I say. 'I definitely saw some strange activity.'

'And are you impressed with tonight?'

'Very much so,' I say.

'Well, back to David,' Yvette says. 'We're all very impressed here.'

TO: Pastor James Worth
FROM: Will Storr
SUBJECT: Interview request

Dear Pastor Worth

I am currently doing some research into ghosts and the afterlife. I found a webpage that refers to you as an exorcist. In the UK, it is very difficult to get clergy to speak openly about this important subject. I was wondering if it would be possible for me to travel to the US to speak to you.

Many thanks.
Will

TO: Will Storr
FROM: Pastor James Worth
SUBJECT: RE: Interview request

Exorcism is not for entertainment or for gain.
This is not a toy.

Read Matthew Chapter 10.

In His grace,
James <X><

TO: Pastor James Worth
FROM: Will Storr
SUBJECT: RE RE: Interview request

Dear Pastor Worth

My aim is to show how dangerous divination can be, and impress on people the risks they are taking when they use Ouija boards etc. I would have thought you would have been willing to help me in this.

Will Storr

TO: Will Storr
FROM: Pastor James Worth
SUBJECT: RE RE RE: Interview request

You are playing with fire.

In His grace,
James <X><

7

'All I ask is that you put your life in my hands'

I'm a tongue-twister away from breaking someone's jaw. It's just past four in the morning and I'm sitting, cross-legged, on the floor of a cold Welsh museum. There are twelve of us, holding hands in a circle, uncomfortable and awkward like a human crown of thorns. In the middle, a man called Tim is sitting on a small wooden chair, reading out tongue-twisters in a squeaky voice. We've been told by the enigmatic leader of the group, who likes to be known as 'the Founder', to concentrate on the chair and to try and make it levitate with the power of our thoughts.

'Six sick slick slim sycamore saplings,' says Tim.

I'm not sure what the rest of these people are thinking ...

'What time does the wristwatch strap shop shut?'

... but I'm thinking that I'm freezing and exhausted and I've been sitting on my legs for twenty-five minutes now ...

'Vincent vowed vengeance very vehemently.'

... and if this so-called 'Tongue-Twisting Magic Chair Experiment' doesn't end in a minute ...

'Chris Cringle carefully crunched on candy canes.'

... I'm going to levitate that chair out of the window with the power of my fucking foot.

Just as my anger is peaking, the strip light on the far wall, which throws its clinical white light over a mural of a Spitfire that commemorates this place's role as a wartime RAF base, flickers off. There's nobody by the switch and we've been in and out of this room for hours – the light has been fine. We sit for a few seconds in frightened darkness. The strip light flickers on again.

'Just think, that sphinx has a sphincter that stinks.'

I try to distract myself from the unpleasant present by mulling over my night at *Most Haunted*. I suppose the reality is that TV shows are expensive to make and complicated to organise and their makers are likely to leave as little to chance as possible. But, that said, leaving Derek Acorah's crazy displays aside, I don't think there was anything scripted about the batteries on the crew's lights suddenly draining. But is it really possible that spirits are somehow using the electricity? Could there have been invisible ghosts in St Nicholas's graveyard, sucking like feeder fish from high-powered lamps?

'Don't pamper damp scamp tramps that camp under ramp lamps.'

It was on the set of *Most Haunted* that I first heard about the Founder, who also goes by the name of Dave Vee. There was a rumour that he caused a small sensation by walking off the set of *I'm Famous and Frightened*, another Living TV ghost show – this one featuring various down-on-their-luck celebrities locked inside Chillingham Castle in Northumberland. Vee was hired as the resident paranormal expert and fled the show in a rage, fearing for his paranormal reputation. He was appalled, I heard, by its flippant treatment of a serious subject. Vee had the courage to resist the many and magnificent seductions of television and so, when I found out that he ran his own research organisation, Ghosts-UK, I was happy to pay the £35 joining fee and sign up for the next investigation.

I was picked up yesterday morning outside London Bridge station by a forty-two-year-old ex-soldier called George. He was wearing a black shirt that had been ironed with military precision and had the group's logo stitched into its chest and sleeves. George is a 'five star' member of Ghosts-UK. He is the Associate Member Investigation Co-ordinator and, if you access his details on the site, you'll find out that 'Big G', as he is known, is also a

Chat-Room Administrator and that his stars are coloured gold. This sort of information, I quickly discovered, is of considerable importance to Ghost-UK personnel.

On our way to Wales, we picked up George's friend Sarah. For the next six hours I tried to get some pre-vigil sleep and, as we motored along, was spinning in and out of consciousness. During periods of lucidity, as we bombed up the motorway, I heard snatches of complex tales of internecine fighting, some remarkable free-form bitching, endless debate about the worthiness or otherwise (it was mostly otherwise) of various members' star-ratings and much discussion of regular inhabitants of the chat-room. At one point, I even heard a white witch's whiteness being called into question.

Eventually, though, I'm roused from my nauseous, disturbed and bumpy sleep by the squeak and sit of the car pulling up, and by George saying, 'I said to him, "Change the fucking job title on the fucking site or I'm fucking out of here for good."'

We have arrived at Maes Artro, an ex-RAF camp in Gwynedd, North Wales. It's a low huddle of corrugated roofs, mossy bunkers and concrete walled huts that was used in the Second World War as a base-camp for pilots. It's now a museum and every few weeks a 'forties ball' is held in the old bar. People come from all over North Wales to dress up in period costumes, drink heavily and jive to Glenn Miller. During these events, strange anomalies have started to show up in photographs: people in perfect forties outfits and hair-dos. People who nobody knows. Could these mysterious strangers be the spirits of dead dancers, reappearing for post-mortal fun and flirting? G-UK are here to find out.

By early evening, twenty-five members of varying rank have arrived and are sitting in the large, warmly lit canteen. We've all been served a small bowl of thin turkey stew, which proprietor Shirley proudly informed me was 'free', coming, as it did, as part of the £25 fee we've all paid to be here tonight. I eat my dinner alone, George having deserted me to sit with other senior members, all of whom are turned out in packet-fresh G-UK branded shirts, fleeces and raincoats. They eat their meal noisily and discuss an absentee member.

'And she reckons she's a psychic?'

'Bethany's not a bloody psychic.'

'Bethany's about as bloody psychic as this bloody stew.'

Everybody George is talking to is called Steve – there's Margate Steve, Steve PM and Big S Steve. Each Steve is in his late thirties and significantly out of shape, and one of them has upsetting, medieval teeth.

Earlier, when I asked George if the Founder would be joining us, he just laughed at me, as if I were indulging in crazy talk. But, during dinner, a rumour started floating through the group's chatter like a spectral orb – Ghost-UK's celebrated leader, they said, is on his way.

Once we've finished dinner, Big S – a loud Northern Irishman with tattoos on his forearms and a large rectangular head – stands at the end of the hall and shushes the gabbling crowd. It's time for our pre-vigil talk.

'I,' he says, rubbing his hands together, 'am a parapsychologist. As far as I know, none of you lot are ... ' He looks around the group with his eyebrows raised. 'No? No? Thought not. So, listen up. I suppose most of you are here because you've seen Mr Acorah and his band of merry men on TV. Am I right?'

The junior members of Ghost-UK nod.

'Well,' Big S continues, 'as a parapsychologist, I can tell you right now that *Most Haunted* is a load of rubbish. People watch Yvette running around and they think that's what it's all about. It's not. It's about professional, scientific, paranormal research. And to do that research we need some very expensive equipment.'

Big S's preaching reminds me of a church service. Even down to the people. There are a notable number of middle-aged mothers and fathers who, no doubt, relish their time away from the drub and drag of the slowly revolving daily routine. What better way to escape a life spent collecting reward points, ironing socks and worrying about the cracks in your marriage than the idea that, yes, there *is* more to existence? That society and circumstance have not condemned you to forty more years of dwindling this and then nothing.

Big S picks up a camera tripod. 'Does anybody know what this is?'

'A tripod,' says someone.

'That's right, well done. Does anybody know what we use it for?'

Once Big S has told us what we use a tripod for, he warns us about the dangers of static electricity. 'I could get technical,' he says, 'but I won't.'

We then watch him take an inflated balloon out of a plastic bag. He rubs it on his jumper.

'This demonstrates,' he announces, 'that if people are overexcited, that creates static, and if there's static, ideas can get implanted in your brain.'

A ripple of nodding and murmuring moves through the crowd.

'Right, any questions?' he says, looking around the room, before pointing at me with both hands, his eyebrows raised. 'Any questions?'

I shake my head slowly and retreat into my seat as all the faces look at me. Just at that moment, a slight man with round glasses and a black goatee beard walks into the room, holding two heavy flight-cases.

'Well.' For an instant, Big S looks shaken. It's as if he's seen a ghost. He takes a small, subconscious step backwards. 'Erm, as the Founder is now with us, I suppose I should run through the rules. Right,' he shouts, clapping his hands together and composing himself again, 'listen up and listen good, people.'

The Founder places his bags down, walks softly to the back of the room and stands with his thumb and forefinger resting on his beard, watching. He is a completely different species to his apostles. He's thin and pale and moves slowly and fluidly, giving the impression that he's in a constant state of thinking deeply.

'Rule number one,' continues Big S, with his big hands on his big hips. 'Respect and trust the people you work with. You put your lives in their hands and they put their lives in yours. Rule number two. Look out for each other. Build a bond. Get to know that person personally. Rule number three. Safety is paramount. If you do not feel comfortable, do not stay. I'm not telling you not to stay – I would. But then, I'm a parapsychologist. Rule number four. If there's a fire – *do not run*. Apart from anything else, it might not be a fire. Finally, rule number five. You do not leave your team for any reason. Never. If you want to leave your team, tell your team leader.'

As the G-UK members write down Steve's rules in their notepads, I quietly slip to the back of the room to introduce myself to Dave Vee. Although he looks much younger than his forty-five years, he has the air of someone considerably older. He speaks with a polite, clinically precise and cultivated voice and has the confidence you'd expect of a man who has awarded himself the status of seven gold stars on his own website. Throughout the night, however, he maintains a distance between us that is palpable. At times, it's as if he's talking to me through a frosted shower door.

'So, do you do this full-time?' I ask.

'This?' he says. 'Well, what I do is far more involved than this. This is just like a show and tell for interested people. But yes, the majority of my time is spent on paranormal research. I've been doing it for twenty-nine years now. I am also a musician, though.'

'A musician?'

'Yes, I play guitar for Spider Monkey.'

'Spider Monkey?'

'We were big in the eighties,' he says, visibly disappointed. 'We had a couple of hits. Look, we'll talk later, OK? I've got a lot to be getting on with.' He drifts away.

I am assigned to Big S's group for a séance. We troop off towards a hut that houses the fuselage of an old fighter plane. Once we arrive, temperature readings are taken and five of us hold hands in a circle by the plane's tail. The lights have been switched off and fitful flurries of wind and rain are whipping off the north coast and slashing about the flat roof and thin windows of the building. In the dark, I become aware that everybody seems to be breathing deliberately deeply.

'Is this your first time, Will?' says Big S, opposite me between breaths.

'Yes,' I reply. 'I've never done a séance before. Um... they're not dangerous, are they? It's just that I have been warned ... '

'Trust me,' he says. 'All I ask is that you put your life in my hands.'

'Right,' I say.

'OK, everybody,' Big S continues, 'it's just past midnight. There's no such thing as the "witching hour", can I just tell you that? That's a load of bull. OK, good. During the séance, I may ask you to move to the left or to the right. You're going to have to trust me when I say that. I'm doing that to get us closer to the spirit. All I ask everyone to do now is take deep breaths and picture yourself surrounded in a bubble of light.'

I close my eyes, try to shake the image of a disapproving Father Bill from my mind and concentrate hard on the circle casting. This place, I think, could well be haunted. It's a place that was packed with people living in an extreme time. It's seen plenty of premature death. Maybe there *are* sad souls trapped in here, surfing on the air-streams and feeding off the light fittings.

'To the spirits that are in the fabric of this building,' Big S booms, 'I ask that you come forward in love and light. We are not here to harm you. We wish to understand and pass messages on to your loved ones. I ask that you channel your energy through me and me alone. Please come forth and use our collective energies.'

The atmosphere in the room has thickened noticeably and the wind has reared up aggressively. It's battering down on us, as if it is trying to break in through the windows. Big S continues his thunderous summons for another fifteen minutes and the air between the group seems to become tauter and closer. I can almost understand the idea that something could feed off our energy. It's as if we're acting as some sort of dynamo, as if waves are flowing through us and getting stronger and stronger with each revolution. Then I hear at least two people let out small gasps.

'Can one of you just stand back?' says Steve, softly. 'Lucy? Keep going back. Keep going back. Relax, everybody. I promise nothing's going to harm you.'

Our human circle moves two steps backwards. I keep looking over my shoulder at what everybody else seems to see, but, in the darkness, can only make out the fuselage of the plane, like a charcoal outline on black paper.

'To the spirit that may hear my voice,' Steve says, 'I ask that you now step forward. We are fully aware that you are in the presence of this room.'

'It's behind me,' says a scared female voice. 'Can you feel it?'

'Yes,' squeaks another.

'OK, it's fine, Sarah,' Steve says. 'I promise you, I will protect you.'

I scan the fog of mixed blackness desperately.

'Spirit, I ask that you come forward in love and light. It's just behind you, Sarah.'

'Shit.' Sarah's voice is becoming unstable. 'Shit. I want to stop, sorry, I'm really sorry, I want to stop.'

'To the spirit that may hear my voice,' says Steve, 'I ask that you now stand back. We are grateful for the opportunity that you have bestowed upon us to communicate with you. I now ask that you stand back in love and light ... '

But something's gone wrong.

There's a heavy, gripped silence.

'It's not standing back,' Steve says. 'It's not standing back.'

'Oh, God,' says Sarah. 'Shit!'

'To the spirit that may hear my voice, we ask that you now go back from whence you came. With love and light, we are very, very grateful. Thank you, spirit, thank you. Please stand back.

Thank you. Please, please, please stand back. OK, everybody, it's fine. Let's release.'

We let go of each other's hands.

'Did you see him?' says Big S with a torch shining up on his Easter Island face.

'Yeah,' says everybody.

'Um … I didn't,' I say. I feel like I'm ruining the party, like I've just put my cigarette out in the middle of the wedding cake.

'There was this gentleman standing right there,' says Big S, aiming his torch at the wall.

'And you all saw this guy?' I say.

'Everyone seen it and felt it, didn't we? He was six foot. And he would not stand back.'

Disappointed, I decide to go and find George. I leave the hut, hurtle out through the rowdy weather and step into the bustling canteen. I find him sitting with a middle-aged woman who's got bunches in her hair. They're flicking through a photo album.

'That's an orb that Dave caught at Chingle Hall,' he says.

He flips the page.

'Look at that,' he says. 'That's us when we met Maurice Grosse. He's the most famous ghosthunter in the world.'

A few days ago, Maurice Grosse phoned me. I'd written to him, asking if we could meet and talk about the Enfield poltergeist case. I want to tell George about this, but have to stop myself. I can sense that, somehow, our relationship has taken on a weird sort of master and servant dynamic. I fear that my appointment with the 'most famous ghosthunter in the world' will damage the complex social-hierarchal matrix that's formed between us, like fragile strings of caramelised sugar.

Instead, I walk over to the Founder who is standing up, demonstrating dowsing rods to the museum owner's family. Shirley is watching with her young son and elderly mother, who are sitting around him in a huddle. Unlike the Ghost Club's versions, the sections of these rods that sit in the hand have been slipped into white plastic tubes. Whilst this means that they cannot be manipulated by subconscious muscle movement, it also makes them very loose, and they can slip about easily, by mistake. I sit down with the family.

'Spirit,' says Dave. 'Show me a no.'

We wait. Eventually, one of the rods falls away to the left. Dave doesn't realise it, but from where I'm sitting, I can see that his left hand is the tiniest bit skew-whiff. It's an easy mistake to make, and he's obviously not aware – but because of the weight of the metal rods, and the loose covers that they sit in, it only takes one hand to be slightly wonky and, eventually, the rod will slip. So, it's not a ghost giving replies to his questions, it's gravity.

'OK, then,' the Founder says proudly to Shirley and her rapt relatives. 'This means that when the rods move apart like that, it's a no. But if the spirit crosses them over each other,' he explains, 'it's a yes. Right, let's begin.'

He clears his throat. There's an expectant silence. Shirley looks at her mother and son slightly proprietorially, as if to say, *You just watch this. You're going to be amazed when you witness the miracles I have seen.*

'Spirit,' says Dave, 'are you a lieutenant?'

The three members of the family stare at the rods. Granny and son's mouths are slack and open. We wait. The left rod slips away again.

'That's a no,' he says.

Granny gasps. Shirley's son grips onto her forearm. She pats his fingers, reassuringly.

'OK,' says Dave. 'Are you a squadron leader?'

The left rod slips away again. Shirley shoots her mother an 'I told you so' smile.

'Hmmm,' says Dave, in the manner of a television doctor examining a perplexing X-ray, 'that's another no. Right,' he says, 'are you a pilot?'

I feel like I should tell the Founder that he's mistaken. I feel as if I should tell him that he needs to have his hands absolutely level or his dowsing rods will just carry on falling to the side and appearing to say no. Then it occurs to me that if Dave keeps thinking he's being given no answers to *all* his questions, he's going to end up with a very strange conclusion. I decide to tell him.

Then I decide not to.

The rod slips to the left again.

'Interesting,' says Dave. 'Are you a man?'

It happens again.

'No. OK, then. Are you a woman?'

And again.

'Hmm,' says Dave. I look at him, frowning at his dowsing rods. He's absolutely confounded. 'Not a man and not a woman,' he says to himself. Gran and grandson are almost unable to bear the tension. They shift uncomfortably in their seats. Shirley bites her lip.

There's a quiet pause. Dave remains genuinely confused. And then, he decides.

'We must be in contact with an animal.' He screws up his eyebrows. 'Perhaps a dog.'

With this, the young boy erupts. 'Mum! Mum!' he says. He's bouncing up and down on his seat with one hand gripping Shirley's arm ever tighter. 'It must be Tanya!'

'That's our old dog,' says Shirley, with melting eyes. 'She passed away last year.'

Gran leans into the dowsing rods and says, in loud and deliberate voice, 'Hello, Tanya, it's nice to see you.'

'Do you have grey hair?' says the boy.

'We'll see,' says Dave.

The left rod slips round again.

'No,' says Dave.

'But ... Mum,' says the boy, quietly and upset. 'Tanya did have grey hair.'

A defensive look flashes across Mum's face. 'Tanya's hair was all sorts of colours,' she snaps.

'Perhaps she's trying to confuse us,' says Dave.

Just then, a man wearing G-UK branded trousers shouts, 'Can we have you all in the bar, please? It's time for the Tongue-Twisting Magic Chair Experiment.'

An hour later, I'm aching, miserable and sat in a corner of the canteen with Dave Vee. Most of the members are asleep in the room, sprawled out on tables and benches and curled up in corners. Somewhere nearby, Big S is snoring, monumentally.

'What exactly was that all about?' I say, trying to get the blood back into my legs.

Dave is earnest, calm and cautious and is wearing a black fleece with the word 'Founder' stitched into it.

'As much as anything else, that was a psychological experiment,' he says. 'It's interesting to see how people respond. Some people will become happy, some people will become angry, some people will become upset.'

'I wonder,' I say, above the sound of his follower's nocturnal grunting, 'if spending all this time in scary places has an effect on you. Do you ever have trouble sleeping?'

'I do have trouble sleeping, yes,' he says, looking at his feet and fingering his goatee beard. 'That came about because of an investigation I performed which was quite unusual.'

One day, he tells me, a man from Oxford phoned him up and asked for help with a ghost that had sprouted in his house. Dave and his team looked into it, and after a short amount of time, he says, they stopped investigating the house and instead, started an investigation into the owner.

'The owner?' I say.

'He turned out to be a vampire. That's why I don't sleep very well. A lot of strange things happened after that.'

'Like what?' I say.

'I had two investigators working on that case. The first left for work one morning, and just as she walked through the gate, her house blew up. They've never been able to find a reason for it.

The second one, she got out of her car, walked away, probably thirty or forty yards, and her car exploded. They never found a reason for that, either.'

'Wait a minute,' I say, as a swarm of questions fly into my head and confuse me for a moment, 'how did you work out that this man was a vampire?'

'Through a process of elimination,' Dave says. 'We do very stringent research and put all of that information into a database.'

'And your database told you that he was a vampire?' I say.

'Yes. And there were other very, very strange things. This gentleman, who was unemployed, was never available during the daylight hours. His wife would have to make frequent trips to the hospital for blood transfusions and she was always bitten pretty badly on the neck. As was the pet dog. And there also was a photograph of him on the wall. We had it sent away and it came back as a photograph from the 1800s.'

'So he was over a hundred years old?'

'No,' he says, 'he'd been alive since the thirteenth century. But, if you don't mind, I don't want to say too much else because things go awry when I talk about this.'

'Can you just tell me,' I say, 'was he evil?'

'He was,' Dave says, with his peculiar stillness, 'but I think he'd mellowed through time.'

I ask Dave if he believes, as Lou does, that there is a genuine force of evil at work in the world.

He nods, almost imperceptibly. 'I do,' he says, 'and that wasn't the first time I've battled with those forces. The first time was a place in Hampshire that I was actually living in. It was a shared house and, one night, this gentleman decided to run off. We heard the sound of all these suitcases being thrown down the stairs and the front door slamming and that was it, he was never seen again. It turned out that he'd been trying to learn how to be a black witch and cited incantations from a very old curse book. The following day there was so much activity it was amazing. Luckily, I had a vicar down the road who was a friend. He went up to the room, popped his head round the door. As his head came back, the door was rapping – *doof-doof-doof* – like that, and he said, "You don't want to go in there." But being the inquisitive sort of person that I am, I went to have a look. I saw this demonic figure. I'd never seen anything like it. It was stuck halfway in the wall, because he'd only got the curse half right.'

'And did it have ... horns?' I say.

'Yes. And it had red eyes and a goatee beard and it was hoofed.'

'And was it muscular?'

'Yes, very much so. Now, as all this was happening,' Dave continues, 'I had a strange visitation in the night. When I told my vicar friend what happened, he showed me some mythical illustrations and asked me to identify what I'd seen. When I showed him, he said, "You're very lucky. This is a guardian. You must be honoured. Not many people have seen this."'

I ask Dave Vee if he considers himself to be a sort of 'chosen one'.

'Well, this thing has only shown itself to people of significant greatness, one being a pope in the fourteenth century and the other being Jesus. And that's probably about it. So, yes,' he says, peering at me intently through his small, round glasses, 'I do consider myself to have been chosen.'

8

'Turn the light off, bitch'

The night outside my window is freezing and still, and the room that I'm sitting in is empty, except for me, my desk and all of the mysteries of the universe. A small lamp casts a net of light over my notes, my computer and the hot pot of tea that's sending up a trail of steam that rises and curls and becomes, right in front of my eyes, completely invisible. All I can hear is the hum of the fan in the back of my computer and the occasional distant splashing sound of my girlfriend Farrah in the bath. For some reason, I'm uneasy tonight. The ghosts are obviously soaking into my subconscious. I swallow and look around me nervously, into the corners of the ceiling and the edges of the doorframe.

I pick up the tape player and slide a cassette in. It's got a white sticker on it with the initials COTC spidered along it in blue biro. It's the recording I made a couple of days ago, when I went to visit a paranormal organisation called Children of the City who claim to be experts in EVP.

I pour myself a cup of tea and press play on the tape. A voice comes out of the speakers. It sounds tiny and comical in the looming semi-dark room.

'We've got some of the best EVPs in the world,' it says.

It's the voice of Stacey – a co-founder of COTC. I lean back and, as I do, the old wooden kitchen chair I'm sitting on creaks angrily. I try to squeeze the paranoia out of my eyes and drift back to the scene of a couple of nights ago …

I'm in the lounge of Stacey's small, modern-brick semi in the village of Tarring near the West Sussex coast. There's a seating area, with a dining table behind it, and a rubble-strewn desk with a computer on it in the corner. The room is in a state of morose unkemptness. Its corners are chipped, its carpet unloved and there are crumbs. A clinically depressed cat is lying on the sofa with its tail in an ashtray. Stacey is sat opposite me over the table, smoking a cigarette.

When I overheard some members of the Ghost Club mention Children of the City in hushed and slightly reproachful tones, I decided to look up their website out of curiosity. It turned out that they have a large and impressive EVP collection on it. When I ask her about them, Stacey turns to face her desk, and plays me some of her choice cuts.

'Get out,' says a voice from Stacey's PC.

Like Lou's, they're shrouded in static. But these EVP are much clearer than the demonologist's. You can actually hear what these voices are saying.

'Turn the light off, bitch,' says a ghost.

My host turns back to me and, as she does, she releases a small, involuntary groan. Stacey's large size means that even this minor manoeuvre is something of an effort.

'I mean, that's scary shit, really, isn't it?' she says, and takes a puff on her cigarette. Above Stacey's desk, there's a mirror that has a large blue sticker on it. It reads 'I Am A Goddess'.

'You're sitting in a haunted house now, you know,' she says. 'We get all sorts: bangs, rappings ... '

The goddess stubs her cigarette out with one powerful drive, pushes her sleeves up, leans on her elbows and looks at me with her face. Smoke comes out of her nose.

'Have you ever had a ghost in this room?' I ask.

'Yeah, there's one in the cupboard,' she says, motioning past my shoulder.

I turn and look. There's a black-painted door in the wall that leads to an under-stair space. I look at Stacey, startled, through her smoke.

'So, if I open that cupboard door, right now,' I say, 'will I see a ghost?'

'No, it's more of an energy,' she says, absentmindedly picking up the corner of her box of Lambert and Butler with her thumb and forefinger. 'I'm clairvoyant and even I can't actually see his face. A lot of people feel him, though. I call him my hermit. He just sits in the cupboard. I've no idea why. I just leave him there. Sandy's clairvoyant, too, so she sees him sometimes, don't you, Sandy?'

Sandy's also a member of COTC. She's here to babysit Stacey's two children while we're out in the local graveyard. She's slumped next to the sad cat on the sofa and gives a small nod, before carrying on with her fag in silence.

I turn back to Stacey and ask her when she realised she was clairvoyant.

'Probably when I was about eight or nine,' she says.

'What exactly was it that you noticed?'

'Well, that I was seeing things other people weren't seeing,' she says.

'What things?'

'Dead people,' she laughs, self-consciously. 'Dead people. Walking around.'

'Do you see them a lot?'

'Yeah, all the time,' she says, in that unexpectedly unshowy, matter-of-fact manner that I'm still not getting used to.

'And do you both see things at the same time?' I ask, turning back to Sandy.

'Yes,' Stacey answers for her, 'and my children do as well.'

They say that children, like animals, are able to see ghosts much more readily than adults. Kids are more instinctual. They're not conditioned into seeing and hearing only the things that experience and society has told them they're *supposed* to be seeing and hearing. They don't automatically filter out the anomalies like an adult, who has a fine-tuned, hard-wired, super-efficient grown-up life-computer in his skull. It's like the odd phenomenon of invisible friends. Many otherwise normal youngsters appear utterly convinced that they have companions that nobody else can see. Some paranormal experts are convinced that these 'invisible friends' are, in fact, ghosts.

'So, both your children see these ghosts, too?' I say. 'Is this like the whole "invisible friends" thing?'

'No!' Stacey says. She laughs sarcastically and shakes her head in disbelief. 'They're not invisible!'

As Stacey unsheathes another cigarette, a tapping sound comes from the wall. I jump, and look to Stacey for reassurance.

'Oh, that's nothing,' she says with half a smile. 'Just the wind or something. Honestly, you'll know the real thing when it happens.'

'Shall we go?' I say, and the sound on my tape stops with a static bump, and starts again.

Durrington Cemetery is a short drive from Stacey's house. It's dark, flinty-cold and unsettled in this place. All around me, gravestones jut out of the earth, each one an anchor plunging out from the surface, the only thing keeping the presence of their drowned owners moored in the world of the living.

'How does this work, then?' My voice sounds smaller, a little carbonised on this portion of the tape. We're holding our Dictaphones out on the flat of our palms in front of us as we walk. This is to ensure that nothing accidentally brushes up against the microphone. Stacey also says that she instructs members of COTC to speak loud and clear when they're after EVP so there can be no dispute as to the source of the sound.

She goes on to tell me that if you blow a dog whistle you'll barely hear its high-pitched squeal. But, if you blow onto a recording Dictaphone (preferably a modern digital one) and play it back, you'll hear the sound much more clearly. This proves, she says, that the machines pick up frequencies that are outside the range of the human ear. And ghosts can communicate on these frequencies. Stacey has yet more evidence that EVP aren't, like some claim, stray radio waves. For a start, if these are the voices of radio DJs, they should rattle on and on and on interminably until you feel broken by the whole pointless tedium of it all, like a real DJ does. Stacey has also recorded voices telling her to 'turn the light off' when she had just turned it on, ordered her to 'get out' when she had just entered a room and, best of all, cried 'behind you' when she'd temporarily lost her partner on an investigation. On that occasion, a nearby COTC member had an EMF meter that, she said, went beserk the moment that the EVP was recorded. And, yes – as it turned out, he was behind her.

'Shall we ask some questions, then?' I say.

'Sometimes it's better just to chat,' Stacey says. 'A lot of the time, spirits try to join in the conversation.'

'OK,' I say, and think about how I'm going to frame the next question. When I overheard the Ghost Club people discussing Stacey's group, I got the impression that they thought COTC were a bit dodgy, that they might have been touched by the sinister velvet finger of the occult. Many supernaturalists, including the Church, believe that some haunting situations are linked to the devil. Ghosts, they think, are evil entities that often arrive when people conjure them up for fun using divination. Some sinister groups try to harness their power for their own benefit, as is possible at Andrew Green's Montpelier Road case. It's this shady business that Father Bill and Lou warned me about. And, keen to find out more, my curiosity was piqued when I clicked onto COTC's website and noticed that their logo is a pentagram, which, famously, is the corporate branding of the Satanist. We walk for a while in silence as I try to think up a way of putting it as diplomatically as possible.

Eventually, I just say, 'Are you involved with the occult?'

There is a tense pause.

'Yes,' says Stacey.

I look down at my feet walking along the grass-verged path that weaves around the coffin-beds. I stay quiet, not wanting to say anything that might draw attention to her candour. After a silence, she admits, 'We do have a branch of COTC that centres on the occult. It's led by Charles Walker, who's an expert on that side of things. He spends a lot of time down Clapham Woods where there is a lot of very, very bad stuff.'

Stacey goes on to tell me that an ancient and notorious group called Friends of Hecate do black rituals down there that have turned the woods evil.

'But I don't want to talk about them,' she puffs as we walk along, 'because they're quite dangerous.'

'Dangerous?' I say. 'Why?'

'There's lots of government people involved with it. Please, I don't want to talk about them. Charles may agree to speak with you. But I can't.'

'OK,' I say, and we carry on walking through the grim and silent boulevards of Durrington Cemetery with our recorders stretched out in front of us.

'Using your clairvoyant powers,' I say, 'could you sense a spot that might be particularly good for getting some EVP?'

Stacey snorts again and gives me a look. 'Er, we're in a graveyard,' she says.

In my defence, Stephen the Druid did tell me that graveyards shouldn't have any spirits in them because they leave the body at the death site (unless, of course, occultists have been 'summoning up'). But Stacey disagrees. She tells me that souls can get quite attached to their mortal remains and like to keep watch over them.

As we pace along in the darkness, I try to distract myself out of the morbid fug that's settling upon me by mulling over some of the evidence that I've found. I want to work out whether the jigsaw pieces of ghostlore fit comfortably together. So …

There is, all around us, an invisible, parallel world that we go to when we die. To Christians this place is heaven or hell. To others, it's simply another astral plane. Although its existence hasn't been proved, clues to the reality of this invisible world are legion. Modern cameras, equipped with infra-red technology, are able to pick up light anomalies that are individual souls floating about in this realm. Because these souls, or ghosts, were once human beings, they vary in nature and in personality. Some of them remain in places that were significant during their lifetime. Many ghosts try to interact with our physical world by materialising – sometimes with the help of conventional power sources – into mists, shadows, bangs, winds, smells or apparitions. Reports of these happenings go back as far as recorded history and span all cultures and continents. If ghosts have enough energy, they can also manipulate very sensitive tools – like dowsing rods, Ouija boards and pendulum devices. Contact can also be made with the inhabitants of this parallel universe by using digital tape recorders that can pick up voices that are inaudible to humans. Some of the more dangerous ghosts hijack still-corporeal bodies and 'possess' them. Christians call aggressive spirits 'demons' and believe they come from hell. Psychics can interact with ghosts, in a very impressionistic way, as they're more sensitive to existence on this dimension.

As I'm thinking, a sudden breeze blows through us as we turn a corner past a cracked granite mausoleum. I decide to ask Stacey if she ever worries that she's putting herself in harm's way.

She thinks for a minute and then replies between heavy, exhausted breaths. 'People accuse us of dabbling with things and say we don't know what we're doing, but that's just idiotic. We're actually scientists and we're investigating something that isn't known. Without us lot doing it, it's going to stay unknown.'

While Stacey was talking I suddenly and unaccountably got spooked. A lizard chill slithered through my nervous system. I look around me. There's nobody here but me, the Goddess and several thousand sunken corpses. I try to shake the creeps out of me by going 'brrrr' and pretending that I'm just a bit cold.

'Do you know what?' Stacey puffs, breathlessly. 'I am actually getting a feeling about a location. I feel that over there might be a good place to get some EVPs,' she says.

Stacey has spotted a bench.

'The Church has a very clear policy about EVPs,' she says, as we approach the seat. 'They say it's conjuration of the spirit.'

'La-la.'

Suddenly, at my desk, I sit up bolt straight. The old kitchen chair lets loose a wail of wooden protest. I rewind the tape again.

'The Church has a very clear policy about EVPs. They say it's conjuration of the spirit.'

'La-la.'

And again.

'The Church has a very clear policy about EVPs. They say it's conjuration of the spirit.'

'La-la.'

There's the sound of a woman singing on my tape. It's only a couple of notes, but it's there. I stand up and walk around and sit down again. I rewind the tape.

'The Church has a very clear policy about EVPs. They say it's conjuration of the spirit.'

'La-la.'

It's so small that the adult, rational brain would filter it out. It would dismiss it as unimportant auditory information. I only noticed it because I was carefully listening out for background weirdness. I sit still and stare, the shock temporarily fazing me out. The simple fact is, nobody was with us. Nobody was singing. And yet, I have just heard two sung notes on my Dictaphone.

I plug some earphones into the player and listen again. It's even clearer now. I get out of my chair again and walk into the bathroom. My girlfriend is lying under a loose arrangement of bubbles.

'Listen to this.'

Farrah puts the earphones in her ears and I press play.

'Can you hear that?'

'The singing?' she says.

'There was nobody singing,' I say.

'Fuck off!'

'Honestly.'

'That's sent a chill down my spine,' she says. 'Let me listen again.'

9

'I was very upset at what I saw'

I'm worried about Maurice Grosse's daughter. A pretty, sparky young journalist at the *Cardiff Leader*, Janet seems to preoccupy her father during his darkest moments. As soon as her name is mentioned, it's like someone has blown Maurice's candle out. His natural bright feistiness is unexpectedly completely extinguished and his quick, spirited chatter turns weighty and drowned. It's as if he's been suddenly submerged in something oppressive and cold and sad.

'She sent a card to my son on his birthday,' he says, looking into his cup of tea. 'It was the most extraordinary thing. On the front, there's a picture of a girl with her head swathed in bandages. And it said, "I was going to buy you a bottle of toilet water", and when you turn it over, "but the lid fell on my head." And she'd written on it, "And there won't be much of that left soon, either!" with an arrow pointing to the word "head".'

He runs his tongue along his lower lip, clears his throat and glances away.

'It was as if she knew that she was going to die. She was a pillion rider on a motorcycle, you see.'

Janet Grosse died of head injuries after being in a motorbike accident on her brother's birthday, the same day that he got the card – 5 August 1976. A series of events that Maurice considers to be 'very strange' happened around the tragedy that convinced him that the world was filled with supernatural mysteries, and rerouted his life for ever.

The moment I discover this, on a cold morning in the front room of his well-groomed Muswell Hill semi, I begin to worry. Maurice Grosse is an almost mythical figure in ghostly circles and he's a council member of the SPR and chairman of their Spontaneous Cases Sub-Committee. But is his faith, his passion for the supernatural, just the reaction of a devastated and desperate father trying to convince himself that his daughter is, somehow, still with him? Perhaps his thirty years of paranormal research have just been a frantic rebellion against the scientist who, with calm, certain and heartless logic, would insist that his treasured daughter has simply vanished from existence.

'You can imagine what sort of a state we were in.' Maurice pauses to release a long sigh that's gathered deep in his chest. 'She was only twenty-two.'

I nibble nervously on some ginger cake and look around me. The house is immaculate. There's a smart seventies suburbia feel to the décor in here, an illusion that's only broken by a large, modern television and DVD player. On the other side of the settee is a bar, complete with half-drunk bottles of Scotch, Gordon's and a soda squirter. The only clue in this room to Maurice's strange vocation is the enormous moustache that lives underneath his nose. It's exactly the sort of moustache that you'd expect a paranormal legend to wear. It's as neat and preened as his sitting room, as plump as his sofa and if it were any larger, it would probably draw stares.

'So,' he continues, suddenly bucking up a little, 'because of all these things that happened, I thought, well, I don't know. Is this a sign from Janet? So I decided to join the Society for Psychical Research. I thought, if I'm going to study the psychical, I better do it seriously. Ha, ha,' he laughs and looks at me. 'Makes sense.'

There were ten 'coincidences' in all that compelled Maurice to join the SPR (which is, incidentally, the world's oldest and most respected paranormal research organisation, and the only one that counts two prime ministers – Balfour and Gladstone – amongst its alumni). To be honest, a lot of these coincidences, to me sound just like coincidences. The fact that Janet died on her brother's birthday; that Maurice was unexpectedly asked to take part in a Jewish ceremony that required him to enter a state of 'mourning' a few hours before the crash; that his sister-in-law's clock stopped at 4.20 a.m. – 'approximately' twelve hours after she died. Sad, if unsettling, accidents of circumstance. Other occurrences do, however, seem curious. There was a punishing drought during the summer of 1976. One day, Maurice mused to himself that, if Janet really was trying to give him a sign, she'd make it rain. The next morning, he was astounded to find that the roof of his kitchen, which extended into the garden below Janet's bedroom window, was, inexplicably, soaking wet. And yet the drought wasn't to break for several weeks.

With his impressive personal credentials – Maurice is a fiercely intelligent Second World War veteran and a highly successful inventor – the SPR took their newest recruit seriously enough for him to be sent on the next investigation that came in.

I hadn't heard of the 'Enfield poltergeist' when Lou mentioned it back in Philadelphia. But it was headline news back in September 1977, after two *Daily Mirror* reporters and two photographers witnessed some terrifying incidents in the home of the Hodgson family. It was these journalists who called in the SPR, at a complete loss to understand the things that they'd seen.

'As soon as I got there,' Maurice says, settling back in his shirt and slacks, 'I realised that the case was real because the family was in a very bad state. Everybody was in chaos.'

During the next few months Maurice, along with over thirty other witnesses, would see almost every symptom of a poltergeist infestation that has ever been recorded.

But things began slowly.

'When I first got there,' he says, 'nothing happened for a while. And then I experienced Lego pieces flying across the room, and marbles, and the extraordinary thing was, when you picked them up they were hot – which is relevant to poltergeist-type activity. They appeared to come out of nowhere. I was standing in the kitchen and a T-shirt leapt off the table and flew into the other side of the room while I was standing by it. I thought, well, that's good. Now I've really seen something.'

I sit back and listen to him recount his astonishing story. Along with his SPR-assigned partner, a highly experienced poltergeist researcher called Guy Lyon Playfair, Maurice watched the case gathering slowly, like a diabolical black storm. Sofas would levitate and tip over in front of them; beds, tables and chests of drawers would be spun on the spot and flung; small rocks would fly right over the house; Janet and Rose, the two young daughters, would be hurled out of bed; coins and stones would drop out of the air; dogs would bark in the middle of completely dogless rooms. On one particularly frantic day, Maurice and a visiting neighbour called Peggy heard Rose crying out, 'I can't move! It's holding my leg!' They rushed out to find her standing on the staircase on one leg, with the other leg stretched out behind her. Peggy grabbed one small wrist and Maurice grabbed the other and they both pulled as hard as they could. Rose didn't move.

And then there was the knocking. It was loud and could come from several places at once. 'If you go and listen to it in the wall over here,' says Maurice, 'it'll suddenly come from the wall over there'. And (and I find this detail unaccountably chilling) a run of knocks would often fade in, louder and louder, and then, slowly, out again. As well as all that, there would be electrical disturbances, doors would mysteriously slam, faces would appear in reflections in the windows, brand-new batteries would drain instantaneously.

Most of the activity, however, centred around the eleven-year-old (and somewhat coincidentally named) Janet. She would be thrown out of bed almost every night, go into violent trances and even appear to speak for the thing. One night, to test the poltergeist's power, Guy removed everything from Janet's bedroom to see what would happen if it had nothing to throw. Later that evening, the family was disturbed by a violent wrenching sound coming from upstairs. The iron fireplace had been completely pulled out of the wall.

'Janet was seen levitating at one point,' Maurice tells me. 'Flying round the room.'

'She was actually floating?' I say, putting my tea cup down, carefully.

'Horizontally!' he says.

Thursday, 15 December was a big day for Janet. It was both the day that her first period came and when the strangeness in her Enfield council house reached its climax. David Robertson (the assistant to a professor of physics from Birkbeck College who was also studying the case) was preparing an experiment in a bedroom with her, when she started speaking in the guttural voice of the poltergeist.

'Fuck off, you,' it growled. (This bad language was out of character for Janet, but not for the voice. In common with Kathy Ganiel and the cases I found at the British Library, this possessive ghost had a filthy gob on it. Once, it embarrassed both Janet and her mother in a supermarket by telling a passing shopper to, 'Shit off, you old sod.')

So, obediently, Robertson fucked off out of the room and closed the door behind him. Then, he shouted in to Janet to 'just start by bouncing up and down on the bed'. And as soon as he spoke, he heard the bed creak.

Then, Janet gasped, 'I'm being levitated.'

Immediately, Robertson tried to open the door. He couldn't. He could just hear Janet inside, gasping and whimpering quietly. Downstairs, on hearing the panic, Rose dashed next door to fetch the ever-helpful neighbour Peggy. And then, suddenly, the door opened. Janet was lying on the bed.

'I been floating in the air,' she said.

'Everything all right?' said the voice of Peggy, running up the stairs.

'Fuck off, you,' said the voice.

Despite having witnessed the full panoply of phenomena at the house, Peggy was sceptical about this. So, she gave Janet a red Biro and asked her if she could levitate again, but this time draw a line around the light on the ceiling. There was no way she could get that high, she reasoned, without noisily dragging a bed across the floor. So, they left the room again, heard the quiet gasps and groans and then, silence. Eventually, they heard Janet land and say, 'Oh!' They burst into the room.

Janet looked at Peggy. 'I been through the wall,' she said. 'I went in your bedroom. It was all white and it had no windows.'

'Don't be stupid,' said the scientist, 'that room is just like this one, except backwards.'

'I did,' she insisted, 'and I dropped my book.'

So, Peggy trotted next door and upstairs to her bedroom – a place she was sure Janet had never been. And there, in the middle of the floor, was her book, *Fun and Games for Children.*

At the very instant a mystified Peggy was picking the book off her floor, David Robertson was yards away, next door, looking up at a thin red line that had been drawn around the light fitting.

'Now,' says Maurice, abruptly. 'How do you account for that?'

'I don't know,' I say, frowning a little. 'Um … nobody actually *saw* her float, did they?'

'Ah,' says Maurice with a smile. 'That was the next thing. David took a heavy red cushion from the armchair downstairs – it was about as big as this cushion here.' He pats the seat of the sofa next to him. 'He told Janet, "I want you to throw this out of the window, with the window closed."'

But it wasn't Janet who answered. It was the voice.

'All right, David boy,' it growled. 'I'll make it disappear.'

David closed the door, as usual, and when he heard Janet call out, he turned straight back into the room. Immediately he noticed that the cushion had disappeared. As had one of the curtains.

Outside on the street, a moment earlier, a baker's roundsman was delivering loaves. He'd heard rumours of ghostly shenanigans in the area, but had dismissed them as 'bloody rubbish'. As he walked along that Thursday morning, he was startled to notice a red cushion suddenly appear on the Hodgsons' roof. Then, he looked closer and saw Janet at the window.

A few days later, Maurice and Guy took a statement from the shaken baker, which is recounted in Guy's book *This House is Haunted*. In it, he says he saw Janet

> *bobbing up and down, just as if she were bouncing on her bed. Then articles came swiftly across the room towards the window. They were definitely not thrown at the window, as the articles were going round in a circle, hitting the window, and then bouncing off to continue at the same height, in a clockwise direction. If the articles had been thrown, they would have just hit the window and fallen down. At the same time, the curtains were blowing upwards, into the room. The whole episode was very violent and I was very upset at what I saw.*

'He was terrified Janet was going to come out of the window,' says Maurice. 'And then, across the road, there's a lollipop lady from the local junior school. She saw her, too. So, we had two independent witnesses who had nothing to do with this case saw it happening. Here.' Maurice nods towards to the table. 'Have some more cake.'

As I break off a corner of some battenberg, I get an itchy niggle. Janet. Why did she make David Robertson leave the room before she levitated? Elsewhere, Maurice had told me that some things would only happen in Janet's presence. On one occasion, a journalist even claimed to have got a full hoax confession from Rose (although she insisted afterwards that she was just nodding as he 'went on and on and on').

'Maurice,' I say, 'do you think there's any chance that Janet was playing tricks on you?'

His answer surprises me. 'Of course she played tricks!' he roars. 'They're children! It would have been impossible for children not to play around. But it wasn't the same as the real thing.'

'What about the talking?' I ask. 'How could you tell that wasn't faked?'

'At first I thought it was. I said, "Janet, that voice is coming from you." She said, "It's not, it's coming from behind me."

Anyway, we did a test. First of all, I got a microphone at the front of her throat and one on the back of her neck and I found the neck one was producing sound louder than the one on the throat. Next, I made her hold water in her mouth and then taped it up with sticky tape.'

'And the voice was just as clear?' I ask.

'No,' he says, holding a finger up to halt my assumption, 'not quite as clear. It was a bit garbled, but it was speaking. And when I took the tape off, she spat the water out, so she hadn't swallowed it. Well, with water in your mouth and your mouth taped over, how the hell can you speak? I don't think there's any conjurer in the world can do that.'

Further tests, with a Laryngogram, revealed that the sound was coming from the False Vocal Fold, which you use when you lose your voice. Try and speak with it now, and keep it up for a few minutes. Hurts, doesn't it? According to Maurice, Janet used to speak like that for anywhere up to three hours at a time, without coughing or clearing her throat or it having any noticeable impact on her normal voice.

Guy Playfair's book about the case is at first fascinating, then terrifying and, ultimately, a wholly depressing read. It's the sheer exhaustion of the family that gets you. The relentless dread. The flying out of bed ten times a night. The sleeplessness. The tears. The constant intrusion. The months the entire family spent so scared that they'd sleep together in the same room with the light on. Considering all this, and all of the witnesses, is it really reasonable to suggest, as some have, that Janet and Rose were deliberately causing the mischief?

Perhaps. For a start, Janet *was* caught hoaxing. She hid Maurice's tape player and, on one occasion, bent a spoon. And some credible experts *did* leave convinced that the children were behind the activity.

Certainly, the rational part of me seems desperate to jump on all of this apparently suspicious evidence and use it like a fire extinguisher on all the rest of the burning mysteries.

As I sit here considering it all, I find myself mulling over the similarities between Janet's experience of possession and Kathy Ganiel's. So, I decide to mention the time I spent in Philadelphia with one of America's foremost demonologists.

And Maurice laughs. And laughs. The eighty-five-year-old is so amused he claps his hands in front of his face, his moustache quivering as if it's being tickled.

'Ah, ha, ha!' he says. 'Ho, ha, ha! What a load of rubbish!'

'Really?' I say.

'Don't go near him!' he says, slapping his knee. 'If he says he's a demonologist, don't go near him. The very word! Ha, ha! Why do you think the paranormal's got anything to do with demons?'

'Er,' I say.

'It's just superstition!'

'Well, there was this one woman called Kathy,' I say, desperately trying to keep my neck above the waves of humiliation that are lapping at my jaw, 'and she was using the Ouija board, and –'

'Yes, yes, the usual story. Listen,' he says, fixing me with clear, sparkling eyes, 'I'm not criticising what happened, I'm criticising that he's calling himself a demonologist. I mean, first of all, you've got to prove there's such things as demons. Look, I'm Jewish, but I never mix my religious beliefs with what I'm doing. You just can't do that. You're a scientist.'

I decide to change the subject by bringing up my initial worry about his daughter. As we've talked, I've been struck by Maurice's passion about Enfield, even now, thirty years later. He's ferociously confident on the subject and when we touch on the sceptics, he can get defensive, verging on angry. He's like a preacher defending his faith. Does this betray an emotional heart to what he insists is a purely scientific interest? Has his three-decades-long search for a paranormal reality actually been more about Janet Grosse than Janet Hodgson?

Maurice nods, thoughtfully. 'Well,' he says. 'My scientific interest, yes, stems from the fact that these extraordinary things happened after my daughter's death. It's just that I felt ... well, you can imagine, I was very, very distressed and it was something to grasp hold of. I thought, why are all these different, incredible things happening around her death? There must be something to it. I must take it seriously. But I'm not a Spiritualist. It's not that I want to get in touch with her, or anything. That said, Spiritualists have produced an awful lot of evidence. I've been in séances where very weird things have happened.'

'Really?'

'Yes. Things that I can't possibly explain. Like a table going up to the ceiling and turning upside down. You should get in touch with the Spiritualists' National Union. They'll tell you all about it.'

I wonder what Maurice thinks of the people who, regardless of all his evidence, refuse to believe the story.

'You have to ask yourself,' he says, 'why are people sceptics?' There's a pause. I raise my eyebrows in his direction. 'Because they're human beings and they cannot face the fact that there are unknown things in life. Now, ponder this,' he says, fixing me with a look. 'Science today is busy explaining *how* everything works. But does anybody know *why*? Science only explains how everything obeys laws. But where did those laws originate from? Did they just come out of nowhere? Think about it. But you put this to a scientist and he'll just go, "That's of no interest to me."'

'It makes you quite angry, doesn't it?' I say.

'It infuriates me!' he roars. 'It's absurd! It's absolutely Luddite! These people wouldn't believe it if it happened in front of them. That's the attitude. It's like this obsession they have with hoaxers. Now, I've been to suspected poltergeist cases where it's turned out that they were just very disturbed people, but it's very rare you go to a case and find that it's an actual hoax.'

It's getting late now and, as I prepare to leave, I wonder out loud how Janet is these days.

'She's had four children,' says Maurice, 'but, unfortunately, the eldest boy passed away when he was seventeen. Died in his sleep. Very tragic family.'

'Do you think there's any chance she'd speak to me?'

'Oh, Janet's moved away,' he says. 'She's very publicity shy. You could try, but she's never wanted to talk about it for years, since she was a child.'

As I pack my tape player and brush the crumbs off my notepad, I wonder if Maurice has any advice for me.

'The only pitfalls,' he says, standing up to let me out, 'are the people that you could be meeting. Be very careful of people who say, "Don't touch that, it's evil." And be careful of these bloody local groups who say they're experts in this sort of thing. They're not. Most of all, though, you must keep a level head. I'm an old hand at this – trust me.'

As his front door opens out onto a soberingly standard north London evening that's turning quickly to drizzle, it strikes me that not once has Maurice mentioned being scared.

'Scared?' he says. 'What's to be scared of? I'm not into bloody demonology.'

'Hello, is that Janet?'

'No.'

'Is Janet there, please?'

'OK.'

...

'Hello?'

'Hello, Janet, my name's Will. I hope you don't mind me calling. I'm doing some research about ghosts and Maurice Grosse gave me your number. I was wondering if it would be at all possible to speak to me about your experiences.'

'Oh.'

'I really would appreciate it, Janet. It would be fantastic to meet you.'

'Well, my mother passed away not long ago.'

'Oh, I am sorry.'

'And I never wanted to talk about it while she was still alive ... '

'Could we say next week sometime?'

'Um ... '

'Wednesday? Evening?'

'Er ... '

'Sevenish?'

'Er, well, OK.'

'Oh, thanks, Janet. That's great. I'll see you then.'

10

'Open your eyes'

It's difficult to believe that spring is just a foot above my head. Walking through this abandoned brick warren of tunnels and halls has made me feel as though it might never come again. Light and warmth are alien to this place. The sun has never shone down here and you can tell. Its absolute absence creeps, like a slug's trail, across your skin as you walk. It's as if the permanent and unnatural subterranean night has caused a different evolution to take place in this dank and labyrinthine void. Occasionally, when the flickering light of the oil lamp strokes brightly enough against a wall, you can pick out a few words of some puzzling and desperate graffiti. It appears as if it's been scrawled in a frenzy. Who'd write it? Who'd come down here? Who'd wander the catacombs of Coalhouse Fort, legendary around this part of the Thames Estuary for its ghosts and violent history? These twisting, witchy corridors regularly frighten their custodians, who have long given up using torches because of the constant, instant battery drain and are forced into using a fleet of rusty hurricane lights instead. I wonder, because I really can't imagine, who'd come down here for fun?

And then I realise that the answer to that question is right behind me, chattering excitedly, its breath rising and merging in cold, quick clouds. The Ghost Club would. Obviously.

I point my camera randomly and shoot. It's the third photo I've taken. I look at the LCD on the back and see an orb. I squint at it for a moment and press delete. Then I notice that the red low-power alert is flashing. My battery was charged, fully, just before I left. It usually lasts for hours. Then, my camera goes dead.

It's a fortnight after my meeting with Maurice Grosse, and I've travelled to the far east of Essex for another Ghost Club vigil. The last time they came, members claimed they witnessed a poker game take place above the entrance of the fort, in a room that no longer has a floor. Other club regulars have spoken to me quietly, over heaving pints of esoteric bitter, of mysterious footsteps that have been heard scurrying through the wet tunnels and a definite and overwhelming presence that lurks in one of the arch-roofed rooms that leads off them. Could it be true? Could the invisible world have sprung a leak down in Coalhouse Fort?

There have been military defences here since 1402, and the building that we're standing in now was built by the Victorians in 1874. It last saw active service in the Second World War, when guns that still sit on the roof tried to pick off German aircraft that were on their way to drop their bombs all over London. These days, the oil tankers that slide out to the North Sea on this slow, wide stretch of Thames move silently and unwittingly over plane-wrecks and corpses that still litter the riverbed.

By now, our group has reached Room 24, an old weapons storage space and the epicentre of much of the phantasmic trouble.

'For those that don't know,' says a man in a big black leather coat, loud enough to silence the chatter, 'a word of warning.'

He pauses, enjoying the moment that he's commanded.

'I,' he says, 'am a Trance Medium.'

There are at least twenty of us standing in a wide, wall-hugging circle. Steve has carefully gelled big-brushed hair, a beer-podgy face and the loudly confident tone of a privately educated P.R. executive trying to make himself heard in a singles bar.

'Things tend to talk directly through me,' he says. 'I may appear to be in pain or suffering. I am fine. Whatever's with me might be in pain, but don't worry. I've been doing this for quite a few years and I've never had any trouble with it yet. Right, if everyone can hold hands and breathe deeply and relax.'

We grasp each other's hands and begin to breathe as one. The light from the hurricane lamp in the middle of the floor is spread thinly in the powerful darkness of the room and the faces of this unwieldy investigating team appear boggle-eyed in the shadows. Above the rest of us, Steve's breathing can be heard, pulling in and pushing out, getting gradually louder and higher. And then, he roars.

'Aaaaaaarrrrrrrggggggghhhhhhhhhhhh,' he says.

A self-conscious fear swells up in the room and bullies its silent occupants into watching him with undivided attention.

Steve breathes in again.

'Eeerrrrrggghhhuuuuuuuuhhhhhhh,' he says, on the out-breath.

Eyes get wider and grips get stronger. Steve's snarls become even more ferocious. Penny, our group leader, decides to take charge.

'Will you come forward and tell us your name?' she says.

'Rarrrrrrrgggggggghhhhhharrrrrrr!' says Steve, even louder than before.

'Please be gentle on the medium,' says Penny.

'Hoooooooaaaaaarrrrrrgghhhhh,' says Steve, louder still, his neck jutting out and his head moving from side to side, like a riled T-Rex in an old Hollywood film.

'Will you come forward and tell us your name?'

'Weee arrrre heeerrreeee,' says Steve, his words long, gruff and gravelly.

'Who is here?' says Penny.

'Weeeee aaaarrrrreeeeee.'

'Who is here?'

'Aaaaaaaalllll ooofff uuuusssss.'

'Who are you?'

'Heeee iiiiis stoppppinggg ussss.'

'Can he speak to us?'

'R o o o o o o o o o a a a a a a a a a a a a a a a r r r r r r r r r r r ! Reaaaaaaaaaiiiiiiiirrrrrrrr!' says Steve.

'Is he afraid of Malcolm?'

Malcolm is one of the workers at Coalhouse Fort. He's six foot seven, wears tight, dirty trousers and a tatty leather waistcoat. He makes baffling jokes and laughs at them while looking at you, and he carries his smoking paraphernalia in a small Tupperware box. In truth, Malcolm is scary on about six different levels. But I can't imagine why any *ghost* would be frightened of him.

'I'll tell you what,' says Malcolm.

He's had an idea. We watch him walk out of the room whilst the possessed trance medium grizzles and spits. Almost immediately, Malcolm's scheme works. As soon as he's hidden himself round the corner, the leader of the nastiness, the one who is 'holding the others here', steps forward in Steve's trance. Its arrival is heralded by an almighty fifteen-second roar that builds to a dramatic, boiling, animal climax, with Steve's neck arched upwards and his mouth stretched open, aiming his vociferous rage at the dripping ceiling. Then, he collapses. If it wasn't for the people either side of him, holding him up by the armpits, Steve would have landed on his face. But they're struggling. The force of his tormented, seized body is dragging them around and they keep having to re-step to stop themselves from being pulled over.

'Tell us why you hold these people,' says Penny.

'Ttthhheeeeeeyyyy aaaarrrrrrreeeeeeeee miiiinnnneee! Rrrrrraaaarrrrggghhhhh!! Ttthhheeeeyyy aaarrreeee miiiiiiine!'

'Why are you afraid of Malcolm?'

'Rrrraaarrrrggghheeeeiiiiiiiiaaaaaaaarrrrrggggghheeeeeeeee,' says Steve.

'OK, it's time to step back,' says Penny, nervously. 'Come back to us, Steve. Come back to the light.'

'Hoooooooaaaaaaarrrgghhhh!'

'Can somebody get Malcolm back, please?' says Penny.

'Rraaaaarrgghhhh!'

Malcolm walks back in from round the corner and approaches the banjaxed medium.

'Stand up!' he shouts.

Nothing happens. The Ghost Club stares at Malcolm. He considers the situation carefully for a moment. Then he decides on a change of tactics.

'Stand up,' he whispers.

The members peer down at Steve to see if it's worked. He's still grumbling and murmuring in a diabolical fashion, hanging by the armpits off the people either side of him. Malcolm reflects on the problem for a little longer.

Then, he has another idea. This time, he thinks, he'll growl his instruction in the manner of the beast.

'Sssstttttaaaannnndddd uuuuuupppppp,' he growls.

The rapt and anxious faces blink out from the shadows at Malcolm, who, on realising that saying 'stand up' is not going to work no matter how he says it, has crouched down and gripped Steve's jaw in his hand. As I watch all of this, a small panic begins to well up inside of me.

'Heeeaaarrrrrrr me,' Malcolm growls. 'Hear ME! Come back now. Open your eyes. Open your eyes. See me. See ME!'

Lately, I've been losing my faith.

'Opeeeennn yooouuurrrr eyesssssss,' Malcolm growls.

What if everything I've found over the last few months *has* just been dust, insects and undiagnosed mental illness?

'I'm trying,' Steve whimpers.

'OPEN YOUR EYES!' Malcolm demands.

I've been thinking ... what if there *is* no such thing as ghosts?

'God,' says Steve, his human voice restored. 'I've suddenly realised you're all stood around me.'

He smiles with faux embarrassment at all the eyes that are beaming onto him.

'What was I saying?' he says.

Someone in the group steps forward, nervously.

'Gain control,' he says. 'Getting control. Getting stronger.'

'The people he's holding are still here,' says someone else.

'What he actually said is, they're his,' says yet another member. 'They're his and he keeps them here because he uses their power.'

'But he didn't actually say that, did he?' I say.

'One of the words actually sounded like "revenge",' says a woman.

'That's what I heard as well,' says another.

'I heard "eyes",' says another member.

'Yeah,' says a man in a woolly Russian hat, 'I caught that bit. He doesn't like Malcolm's eyes. Yes,' he considers, 'there's definitely something about your eyes that he doesn't like.'

As the Ghost Club members crowd around Steve and pat him on the back, I find myself drifting back. I can't join in. Up until the other day, my faith in ghosts came in floods. But now, something has switched. It wasn't a rational thing, based on a careful weighing of the facts. It was an instinctual shift. It happened after my visit to Maurice Grosse's house. What he told me was so incredible that it forced me into a corner. If I was to believe the Enfield story, it would mean that I'd have to sign up, completely, to being a ghost believer of the most fervent kind. Because what Maurice told me overrules all the other arguments: ghosts exist and they know exactly what they're doing. What happened there must have been the work of an outside entity. For a start, it spoke. It threw Lego at press photographers and pulled Rose's leg and broke fundamental laws of physics on request. This wasn't a stored-up tape recording or an obscure psychological effect. It was active. It was behaving. It was on purpose. It bullied the Hodgson family and it smashed up their house.

And yet, something in my brain has shut down. I was willing to go along with all this ghost stuff when it was just flashes and touches and sounds. But when I was forced to go the whole distance, to believe in this crowded, invisible, everywhere world, my mind had a sit-down strike. It's like R.E. A-level all over again.

As I drift further away from the Ghost Club's excitable

throng, I catch the eye of someone else, a thirty-something man with dramatically atypical hair who looks similarly lost in thought. We look at each other for a moment. The beginning of a smile tingles at the sides of his mouth. We can tell what we're thinking. I follow him out of the room and back into the tunnel.

'I wasn't really 100 per cent sure about all that,' I begin, cautiously.

'Yes,' he says, walking briskly in front of me. 'The more of these things I go to, the more I feel … ' He's struggling to say the words. 'It's all … '

'What did you think of the trance medium?' I ask.

'Do you want the truth?' he says, turning as he walks and making full face contact for the first time.

'Yes,' I reply, as we round the corner out of the tunnel system into the warm, copper-scented spring night.

'Utter crap,' he says, finally. He smiles broadly with the relief of it all. 'Total bullshit.' He looks like a man exorcised.

We walk up the stairs, sit down on the roof next to a massive gun and introduce ourselves. Philip Hutchinson is, it turns out, a very senior Ghost Club member. He's on the board. He also makes a living out of spooks, taking highly popular ghost walks around Guildford and London. Behind us, the turrets, platforms and chimneys of some vast industrial plant towers over the estuary. Red and green lights flash, streams of white smoke barrel purposefully upwards and a silver chimney aims a brilliant, computer-controlled flame at the heavens. It's every bit as breath-taking as a mythical castle, supreme and intimidating and awesome against the night.

'In all the investigations I've been on – and there's been a lot …' he says.

And then he stops and goes into a deep nod. When his chin reaches its lowest point he just keeps it there and looks, for a moment, at his shoes, as he considers what he's about to say very carefully.

Then, he looks me right in the eye and says, 'I don't think anything truly paranormal has ever happened.'

There's a silence. Suddenly, a conspiratorial, confessional fever overcomes me. It's mischievous and warm and infectious, and I feel a surge of relief at suddenly being able to give my worries a voice. I move closer and start to tell him about my first Ghost Club outing to Michelham and the dowsing rods that sprung to life in my hands. I tell him that I've been thinking a lot about that night. And I've been having doubts.

Firstly, there was Lance's worry that they'd made contact with a Benedictine monk when Michelham was actually an Augustinian priory. Secondly, there was the Founder's wonky hand in Wales. And thirdly, and most of all, there was the lack of actual researchable facts that were gleaned from Sir Thomas Sackville. We failed to discover what year he was born or how he died. All we knew for sure was that he was dead and angry. So, we had no evidence whatsoever that we were actually in contact with a ghost.

'I've done plenty of dowsing in my time,' Philip begins.

'And have you ever had any results that actually check out factually?' I ask.

'No,' replies Philip, scratching his knee. 'No, not at all.' He looks at me for a moment to gauge my reaction, before continuing. 'And nor has anyone else.'

'It is strange, though,' I say. 'I'm sure that, when I did it, I could really feel a definite pull.'

'Yes, I've been thinking about that,' he says, 'and I think that it's really all due to physics. When you're holding a copper rod in front of you that extends for about a foot, and your hand moves downwards, even just a fraction, the rods will move and the pull of them, the momentum, will actually be quite great. The way you hold those rods, they're effectively quite heavy. And as soon as they start shifting around, the momentum is going to feel like a really strong drag. If I'm totally honest, I think all dowsing rods should come with a spirit level on them.'

I sigh and look down and kick a weed to death with my foot. From deep underneath us, we can just hear the sound of Steve roaring again.

'Do you think all these mediums are just faking it, then?' I ask.

'No, not all of them,' says Philip. 'What's happening down there, it's hugely entertaining, but it's just a show. He's just an attention-seeker.'

'What about someone like Paolo?' I say. Paolo didn't strike me as an attention-seeker at all. He was genuinely terrified in that priory room.

'No, Paolo doesn't fake anything,' says Philip. 'Not at all. Anything that happens he will believe it to be genuine. He's just highly strung, that's all. And he will often go for the paranormal explanation rather than the rational.'

'So you think it's all rubbish, then?' I say.

'No,' says Phil, 'but only because I have actually seen a ghost. I had a 100 per cent certifiable sighting of a woman in a place called Woodchester Mansion in Gloucester on 27 January 2002.'

Philip goes on to tell me that he has two theories on ghosts. The first is the Stone Tape theory, which Stephen the Druid told me about. And the second involves time-slips. Apparently, Einstein's theory of relativity says that time isn't linear, but all twisted up like a ball of wool. So, if we're all barrelling up and down and around and around in these tangled woolly time-strands, it's possible to rub up against a strand from a completely different era, like the sixteenth century, or the twenty-seventh, and when this happens, information can leak through. Thus ghosts.

'This makes me believe that if you can see a historic ghost,' Philip says, 'then they will be able to see you, too.'

I find the time-slip theory difficult to grasp because, surely, when something has happened, it doesn't exist any more. Time isn't like tape from a video recorder that you can rewind and fast-forward, it's like steam from a kettle – once it's gone, it's gone. Mind you, if Einstein reckons this is how it is, I'm not about to argue. But then there is another immediately obvious problem with this theory.

'Hang on a minute,' I say. 'So if ghosts are just people in the past looking back at us through some sort of translucent time-wall, how come we don't see ghosts of the future?'

Phil cocks his eyebrow. 'How do we know that we don't?' he says, smiling in an 'a-ha!' sort of way.

'Because they'd all be wearing space suits,' I say.

'Hmm, well, OK,' says Phil, placing his chin in his hooked forefinger. 'Maybe if the future hasn't yet occurred, we can only see into the past. The past has happened, that's an undisputed fact. So maybe we can only be retrospective.'

He looks at me and thinks a bit more with his lips all bunched up into the middle.

'I've just contradicted myself, haven't I?' he says. 'I need to give this some more thought.'

'I think we both do,' I say, as a long trail of Ghost Club investigators emerges from the depths out of an iron door beneath us.

'Hello, is that Will?'

'Yes, hello. Is that Janet?'

'Yes. I was just ringing to say that I won't be able to speak to you tonight.'

'Oh, right. That's a shame. Can we arrange a new date?'

'No, um ... I'll call you about that.'

'Oh. OK. Well, I hope everything's all right?'

'Thanks, bye.'

11

'I promise you, you'll scare yourself'

Today, the corridors that I'm walking down have seen more agony than Coalhouse Fort has in all of its lifetimes. And yet there's no darkness here. The air is big and dry, footsteps echo purposefully around wide, polished corners and there's light here, constantly. It watches, humming in high, fluorescent tubes and glowing in sad, bedside bulbs as the poor souls that it keeps from the darkness come and go. It's there during the good times, when they leave upright and out of the door and it's there, also, at the bad times when they just flicker off, and their life slips away into the gap between the seconds. But you don't need EMF meters, infra-red cameras or a celebrity medium to find haunted here. And you don't have to fall into a noisy trance to find rage in this place, either. Or turn the lights off to find fear. Just step into the car park and listen to the woman I've just passed who's barking her tears down her mobile phone. There's fear and rage. There's haunted.

I walk quickly, with my head down. I don't like hospitals. They're the waiting rooms for the graveyards – and this is where the grim stuff happens, the terror before the silence. I hurry to a lift with my hands in my pockets, travel two floors up, and am relieved to find myself in a safe, administration part of the building. I relax and enjoy breathing air that smells of carpet panels and photocopying, after the disinfectant tang of the wards downstairs. I find my door, the one with 'Dr Mark Salter' written on it. I knock, and a clean-cut, kind-looking, thirty-something man holds out his hand to greet me. Then, before I've even had the chance to take my cycle helmet off, he says, 'You know, there's no such thing as the supernatural.'

I freeze. After the events of the last few months, this sounds like heresy. Revolution. I take a steadying breath.

Then, Dr Mark says brightly, as if he's just said nothing at all, 'Can I get you a cup of tea?'

I've come to see a shrink. Dr Mark Salter is a psychiatrist who works at the Homerton Hospital in east London and is an expert in delusions of the paranormal. If Dr Mark could meet Lou Gentile, David Vee, Debbie and Trevor and all the rest of them, he wouldn't treat them as professional researchers working on the frontiers of human knowledge. He would treat them as patients. Mental patients. So, I'm here to hear his point of view. I want Dr Mark to speak for the rational part of me that reared up in the black and madness of the Coalhouse Fort tunnels, the voice that protested, a decade ago, against the R.E. teacher who told me that God once pushed her on a swing.

While Dr Mark is out of his office making tea, I check the shelves for supernatural nick-nackery. There's nothing. No candles, crystals, crucifixes, mini-menorahs or pictures of white wolves howling at the moon. There's just a pen-holder, a picture of a family and a big red book called *The Broken Brain*.

Dr Mark returns, sets the cups down carefully and asks, 'Who's she?'

He's looking at the screen of my laptop, which I've set up on the desk. It's frozen at the beginning of the mini-movie that Trevor, the sceptical monsterologist from Avalon Skies, sent me of his girlfriend Debbie under possession.

'That's Debbie,' I say. 'She's a witch.'

The psychiatrist smiles. 'Oh, yeah?' he says.

I press play.

The footage is shot in the blank, reptilian green of an infra-red camera. We're inside a traditional old pub. There are wooden slab tables, grubby glass ashtrays and dusty knuckles of hops hanging down from the ceiling. Debbie is sitting next to Trevor. A shawl is draped over her shoulders and the pits of her eyes glow white in the grainy, inverse light.

'I am George Turnbull,' she hollers into the air. She sounds bloated with rage and as masculine as a fat butcher. Her whole frame is consumed with it. Her neck hurls forwards as she cries out, her lips arch forcefully as they over-enunciate. It's as if someone else has taken over the workings of her mouth and hasn't quite got used to the controls.

'Why are you angry, George Turnbull?' says Trevor. You can see him in the flat greenness, his wide, succulent eyes glistening with worry for his partner. But he presses on dutifully, professionally.

'I got accused and it was not me! I got accused of hiding all the stuff! And it was not me! I will not be accused any more!' she shouts, her eyelids stretched.

Trevor has had enough. He puts his hand on the back of her neck in a cautious, comforting gesture, and says in a firmer voice, 'All right, Debbie. Come forward, Debbie.'

'No!' she shouts, the fury rippling like a layer of boiling blood just under her skin.

'Debbie?' says Trevor.

Everything dives out of view for a few seconds as the camera operator panics.

'Debbie?' says Trevor again. 'Debbie?'

She appears again as the cameraman steadies himself.

'No!' she shouts. Her ligaments are straining, her tendons popping. 'I will not be! I will not be accused!'

Even on this bright and birdsongy afternoon, I can't help but find the footage a little disturbing. I turn to Dr Mark for his reaction.

'Ha!' he says. 'Ha, ha! Great stuff.'

'What do you think?' I say.

'I see this routinely,' he says. 'People get into intense, super-aroused states for whatever reason. This is how someone might look if they'd just found out that their husband has been buggering their six-year-old daughter, or whatever. It's very common. Who's he?' he says, pointing to the now-still image of Trevor, whose arms are clamped in a tight, protective hold over Debbie's shoulders.

'That's Trevor,' I say. 'He's a monsterologist.'

Dr Salter gives me a look and crosses his legs.

'Are you telling me that Debbie's just worked *herself* up into this state?' I ask him.

'Yes, and she's very adept at it, isn't she?' he says, reaching for his tea.

I go for my cup as well and take a moment to consider the shrink. He's wearing comfortable black shoes, discreet wire-rimmed glasses and light-brown semi-casual trousers. Here, I think, is a functional man. A man who's comfortable being one in a crowd. A man whose skull works. But that's not to say he's boring. I don't know it yet, but the two hours I'm about to spend with Dr Mark Salter will prove to be illuminating, compelling and horrifying on the profoundest level imaginable.

'Look at him,' says the doctor, motioning at Trevor. 'The solid, loving, enduring boyfriend whose partner has got this special talent. And it's very important for him, I suspect, to have a girlfriend like this. "I'm not going out with any ordinary woman,"' he says. '"She's a witch."'

'Is she mental, then?'

'Well,' he says, folding his arms, 'she's mentally disturbed. She's aroused, she's hyper-ventilating, she's hurting her vocal chords. I would say that she has the capacity to go into a mildly histrionically sensitive state. And she can turn it on and off at will. Actors can do it. If you talk to an actor just after he's come off stage, he will take four or five seconds before he comes back to who he is.'

That may be so, but I'm convinced that Debbie isn't pretending. Even when you're channel-surfing the TV, the difference between real documentary footage and an actor is instantly obvious. Have you ever seen a *Crimewatch* reconstruction actor do a fear face? Exactly. And there's nothing about Debbie or the majority of the others that made me think, instinctually, that they were acting.

'Oh, she does believe that it's real,' says Dr Mark, 'yes. And it's the same with all these cases.' He turns and lifts the corner of the dossier that I sent him a couple of weeks ago. These are the full, unedited interview transcripts from my meetings with Christopher Tuckett, Dave Vee and Lou Gentile. The cases for the defence. I wanted his psychiatric opinion on whether any of them were lying. Because I remain solidly, up to my hips in concrete, rooted in the opinion that they were all sincere.

'You're right,' he says, 'they were sincere. There's an old phrase – if a patient complains of pain, they're in pain. If a patient tells you they've seen a ghost, they've seen a ghost.'

I lean forwards, towards the psychiatrist. 'So they *have* seen a ghost?'

The psychiatrist sits back on his seat. 'You need to accept that what they're saying, they think is true. So your question is, why do they *think* they've seen a ghost? Well, hallucinations, incorrect memories, confusion of past and present. I mean, these guys who saw the demons, they could be remembering something they saw when they were scared shitless by a science-fiction film their brother was watching when they were five and they shouldn't have stayed up that late.'

But Debbie was not having an incorrect memory. She was demonstrating a different phenomenon and showing different symptoms. Like Stacey, Loping Coyote and Derek Acorah, Debbie is a medium, and mediums – let's be blunt about this – hear voices in their heads.

'You could call it multiple personality if you want,' says Dr Mark. 'We've all got the capacity to have them.'

There's a silence. I put my tea down on the floor between my feet. 'I don't have multiple personalities,' I say.

'Try a little trick for me,' he says. 'How are you going to get home from here?'

'I'm going to cycle,' I say.

'Well, as you're cycling home, imagine a version A and a version B of yourself and have a conversation in your head between them. You know, like: What are you doing tonight? Well, I thought I might write up this interview. Well, you could, but there's some great telly on. I know, but I have got a lot to get through. Carry on like that for an hour and see where you end up. I promise you you'll scare yourself.'

Dr Mark tells me that, before long, the first voice will become more extrovert, more outgoing and prefer art and German techno. The second voice will be quieter, more nervous and like science and South American heavy metal. In other words, they will develop distinct and consistent personalities of their own. Just like a medium and their 'spirit guides'.

'Now, you tell me where that comes from,' he says.

I nod and pick up my tea and I smile silently because the answer is – the answer must be – that both these characters would be a creation of my own mind.

'These people,' he says, motioning with a nod and an eyebrow towards my laptop, 'are erecting other versions of themselves as a way of bolstering up some sort of personal currency with others. You very often find, when you interview a lot of these people, like I do, that the people who have experiences like this need meaning in their lives. They're using it as a way of shoring up some other huge hole.'

So Dr Mark thinks that clairvoyants' magic powers, their walkie-talkies to the dead zone, are just an illusion they've unwittingly created to help shore themselves up in some way. But how does it actually work? Dr Mark's trick shows how you can *willingly* create another personality in your head. But how do you get to a point where there are two different operators working the controls in the cockpit behind your eyes, without you even trying?

'Why aren't you thinking about the planet Jupiter right now?' asks Dr Mark.

'What?'

'Do you know what the planet Jupiter is?' he says.

'Yes.'

'Then why wasn't it in your head just now?'

'Well, because –'

'Because there must be some mechanism that's keeping Jupiter and the other 150,000 words of the English language suppressed, and out of your pre-conscious. And there's a similar mechanism keeping your internal monologue – the voice in your head – simple. Now, supposing those mechanisms go wrong. Did you know that if you go up to the population at random and say, "Do you hear voices?" about eight and a half to ten per cent will say yes?'

So if the bit of your brain that keeps your monologue a monologue develops a problem, your monologue becomes a dialogue. You hear the sound of two of you. But some mediums say they hear *other people's* voices in their heads. Some even claim to have animal spirit guides.

Dr Mark shrugs and glances over my shoulder, out of the window. Somewhere out there, a wailing ambulance is pulling in.

'It can be their own voice,' he says, 'it can be a totally different voice. Often, it's the sound of someone who has emotional significance for them, like their torturer or their abusive uncle. It doesn't go away and it's completely beyond their control. I'll put you in touch with one of my patients, if you like. You'll see how common it is.'

So, what about Vee, Tuckett and Gentile? They aren't hearing sprouted voices. And they're not just seeing flickering lights or fleeting shapes in the darkness, either. Those boys are seeing demons and whirlwinds and looming black shadows. How's that possible? If, like Lou, you think you've seen a ten-foot black shadow with a bowling-ball head in your bedroom, surely you've not mistaken *that* for a poster that's flapping off the wall or a car going past or, indeed, some insects.

'Well, for a start, how old was he when he saw this?' asks Dr Mark.

'Ten.'

'And how old is he now?'

'Mid-thirties,' I say.

'Well, already you're talking about something which has been processed by twenty years of retrospective memory. I mean, I had paranormal experiences when I was eight or nine. I remember waking up and seeing a figure at the foot of my bed.'

'Really?'

'Yeah, you name me a kid who hasn't.'

'Me,' I say. And suddenly, and involuntarily, I hear the voice of a raging Maurice in my head. *These people wouldn't believe it if it happened in front of them.*

Lou's black shadow, as we know, was just the start of it. I pick up the dossier and show the doctor the part where Lou was lying in front of the fire, smelled a foul stench and saw a four-foot demon with 'reddish piercing eyes'.

'Are you going to tell me that he's mistaking the fire for that?' I say.

'No,' says Dr Mark. 'You're right. He's having hallucinatory abnormalities rather than perceptual abnormalities. He's seeing things.'

'But,' I ask, 'what is it that he's seeing?'

'I would think he is reliving a memory. An assembled memory that's taken from various fragments in the past and has something to do with the emotional state that he was in at the time, around the age of ten.'

'But he's not ten now,' I say. 'He's married now. He's an adult.'

Dr Mark sighs, picks up the document and scans through it again.

'Something must have been going on in his life,' he says, shaking his head, 'which, on that particular occasion, for a complexity of reasons that we'll never fully understand, caused something to colour his mind in such a way that in retrospect he's interpreted it as an evil red-eyed lizard.'

But what about Christopher Tuckett? Is he also having a

string of false memories? Dr Mark runs his finger down the interview and reaches a section where I've asked Tuckett how he got used to living in such a terrifying and active place.

'By the end of the first two weeks,' Chris says, 'I was desperate. I wanted this job more than probably anything else I've ever wanted in my entire life, and it became a huge personal battle – do I do the job or do I get scared and go? At the end of the two weeks I got absolutely ruined one evening on wine and it drove me to the point where I was sat on the bottom of the stairs and I was talking to the house. It sounds pretty stupid, I know, but desperation makes you do funny things sometimes.'

Dr Mark points out that, later on, Christopher says that his wife ended up walking out because she couldn't cope with the ghosts any more.

'One thing,' says the psychiatrist, 'that is very, very common in people that seek to convince you of extraordinary phenomena or delusions is that they will often go and select only information that is consistent with their theory. "My wife walked out because she couldn't stand the ghosts." Could it be more likely that his wife walked out because she was married to a man who when he drinks too much, sees ghosts and sits on the stairs talking to the house? It's arbitrary selection. You're putting forward only the facts that suit you.'

The psychiatrist puts the dossier down on his desk and shifts on his seat. He clears his throat and crosses his hands on his lap.

'Listen,' he says. 'Supposing I've come up with this theory that all swans are white. I've written books, toured the world, built my whole career and reputation on my white-swans theory. And then somebody sends me a picture of a black swan. I'm not going to be too happy about that picture, am I? If it gets out, that's my mortgage and my status gone. I'm certainly not going to mention that black swan to anyone, am I? In fact, I might do everything in my power to keep news of that black swan out of circulation.'

The black swan. This reminds me. Since my meeting with Maurice Grosse I've been trying to find a sceptical view on the

Enfield poltergeist case. It's been extremely difficult. And then someone warned me that Maurice had successfully sued a journalist for claiming that the case was untrue. This worried me. It seemed to be extreme behaviour from Grosse. So, I re-read *My House is Haunted*, Playfair's book on the case. It noted an SPR member called Anita Gregory who studied the case for her Ph.D. thesis. She visited the house and left sceptical. So I called the SPR and asked if I could be put in touch with her. The man who answered the phone was the same man who put me in touch with Maurice in the first place.

'Not unless you know a good medium,' he said. 'Mind you,' he laughed, 'you probably do by now, don't you?'

'She's dead?' I said.

'I'm afraid so.'

'Well, can I get hold of her thesis?' I asked.

'Ha! No chance,' said the man. 'And I should warn you, if Maurice Grosse hears you mentioning her name, he'll take your head off.'

So I tracked down John Beloff, the professor who visited the house with Gregory. I wrote to him and requested his opinion. He said: 'I am now eighty-three and my memory is not as good as it was. I do recall, however, that I did make one trip to Enfield and that I failed to witness any notable phenomena. Like Anita Gregory, I suspected that the two little girls involved were up to mischief.' Then, Beloff complains that Maurice 'put Anita Gregory's Ph.D. thesis out of bounds'. I was stunned. Not only had Maurice sued a journalist, he'd also taken action against one of his own fellow researchers.

'Methinks he doth protest too much,' says Dr Mark, when I tell him about my worry. 'It's taking it to extremes, really, isn't it? How did he use a court to prove that it was true?'

'I don't know,' I say. 'I'll have to find out what really happened.'

Before I leave, I want to know what Dr Mark thinks about the vast number of everyday people who have 'seen' ghosts. Not the moustachioed investigators, the paranormal cult leaders, the diamond-studded mediums. I want to know why the ordinary working man or woman might be convinced they've seen a spectral something.

'Because the brain is an explanation-generating machine,' he says. 'Imagine you're driving late at night down the M25. It's approaching dusk and you see something in the distance. You think, bloody hell, there's a cow on the road. You get a bit closer and, after having just seen a documentary on the wildlife of the southern African Congo, you think, oh, it's not a cow, it's an aardvark. Then, as you get closer, you think, oh, no, it's just a piece of blown-out tyre. As further information becomes available to your brain, as you get closer, it keeps generating new explanations. Your brain constantly jumps to all manner of incorrect conclusions. And your state of mind plays a crucial part in determining the meaning you give the incoming data at the time. And by state of mind, I don't just mean feeling cheerful. It's much more complicated than that. It's state of mind, body, anticipation, memory.'

'So, you see something strange.' I say, 'and if you're already inclined to believe in ghosts and if your brain is in a certain mental state, you will think what you've seen is a ghost?'

'At that moment, yes. At that moment, you're drawing on influences from your own past, you're drawing on inferences from what you think is gong to happen in the immediate future, you're drawing inferences coming out from your own internal signals – in other words, stomach tightness, muscle tension, sweatiness, temperature, wind, humidity – all these things that your body is doing. You're also making comparisons between how things are now and how things were five seconds ago. You're constantly updating. You should also consider the micro-environment of a ghostly experience. It's always in that corner, and surprise, surprise, if you check the window ledge, the draughts, the underground passage of water currents, vibrations, strange sounds, the *feng shui* aspects – by which I mean the way the environment happens to be arranged – all those things will be far worse in those particular areas.'

As Dr Mark speaks, I start to think about the moment I was touched by the ghost in the Carvens' house. I was sitting in a darkened room that I had been told was haunted. I had just seen a ghost light float around Mrs Carven's head. I was watching a demonologist speak into a Dictaphone that was visibly, via its flashing light, picking up voices of undead spirits. I was in a state of suffocating terror. And, at that precise point, I was touched.

'Dr Salter,' I say, 'is it possible to be so frightened that you think something's touched you?'

'Oh definitely,' he says. 'Definitely. That fella on the TV does it all the time. Derren Brown. He's brilliant at it. He talks people up until they're grossly over-aroused, then creates this heightened state of expectation and then he suggests. And they fall for it, every Saturday night.'

'So, is that what happened to me when I was touched on the back?'

The psychiatrist smiles. 'It was a puff of air! And you were so over-aroused that a simple puff of air wasn't enough of an explanation for you. Because you'd been pre-conditioned into expecting ghostly phenomena, the first explanation your brain gave to what happened was: I've been touched by a ghost.'

'You know,' says the doctor, as I bend to get my coat, 'human beings have always been desperate to believe in all kinds of supernatural mumbo-jumbo because they are ways of explaining away the most terrifying fact of all. That we are zombies leading meaningless lives.'

'Zombies?'

'Yes. Emotions and free will are just an illusion that we have to stop us blowing our brains out.'

I stop and freeze and listen. Dr James the philosopher said that some people have used the fact that we're *not* zombies to try and prove that we have souls. But is Dr Mark right? Are we just very sophisticated zombies? If so, it's not just religion, ghosts and the afterlife that we use as a comfort blanket when faced with the brutal concept of total death. It's *free will* and our *entire emotional landscape.* Could every decision we make, every feeling we feel, every moral conviction we have, our very sense of self, our personality, our 'soul', our 'consciousness', be just a chimera whipped up by our minds to keep us keeping on?

'Oh, yes,' he says, fiddling with a Biro idly. 'You and I are actually zombies living an automatic life. And we are here for no reason at all.'

To: Milton
From: Will Storr
Subject: Your story

Hi Milton

Thanks for the lift back from the investigation the other night. Much appreciated.

I need to ask a huge favour. You know that off-the-record thing that you told me? Could I possibly ask you to reconsider? Your and your brother's experience is particularly fascinating, for reasons I'll go into if you can spare the time to meet me.

I know your brother doesn't want this coming out. I'm happy not to reveal his name or any clues as to his identity if he wants. If you could have a think about it, I'd be extremely grateful.

Thanks.
Will

12

'They'll build it up and bugger off home'

When he tells the story these days, Charles insists that he wasn't nervous on his way to meet the Satanist. I'm not sure I believe him. He says he didn't have time to worry, that the man who'd called had told him to be there in half an hour, at a crossroads deep in a silent part of the lower woods. This was over thirty years ago now, on a high summer night just like this one. And it would have been dark, during his meeting with the strange caller. It would have been still and eerie and lonely. Charles's thin frame would have moved down the track lightly, his keen eyes darting around the black shapes in the undergrowth that he passed.

Only a handful of people had answered the appeal he'd posted in his local newspaper, for information about the 'mysterious events' in Clapham Woods. And he'd dismissed them all, except this one. This one, he thought, was different. He was well spoken and calm and serious. When Charles reached the crossroads, the place where the caller had told him to go, he found it empty. He looked around. He listened. In that warm, static air, even the trees had barely a murmur. And there was no birdsong at all – but, then, that was normal for this part of the woods. The animals, he already knew, kept themselves well away from this uneasy place. He strode up and down for a time, smoked a cigarette, chewed on his thumbnail. And then, just as he was about to give up and go back to his bike, he heard a voice, somewhere to the left of him.

'Stay where you are,' it said.

Charles froze. The voice, he realised, was coming from the inside of a large bush. He began to turn around.

'Don't turn around,' said the bush. 'Just listen.'

This evening, as he tells me his story, Charles Walker's eyes flit worriedly here and there. And whether he was nervous before the rendezvous or not, he will admit that, when he heard the stern voice coming from the inside of the bush, he was scared.

'I would say I was shaking,' he says to me, his gaze distant with recollection. 'Definitely shaking.'

That moment, all those decades ago, Charles was about to be told something that would set him on a mission that he is still trying to accomplish to this day.

We're at the crossroads now. The tracks are narrow, well walked-in by ramblers, and are surrounded on either side by the brambles and ferns of secondary forest. I'm with Charles, who has been described as 'one of Britain's leading occult researchers and authorities on Satanism'. The fifty-one-year-old is the author of *The Demonic Connection* and an associate of Children of the City – the man Stacey told me about during our evening in Durrington Cemetery. His partner Dave is also here, and it's become clear, from being with them, that they know these paths as well as the local badgers. But, regardless, there's still a tense spring of jumpiness behind their every step. It's the very fact that they know about Clapham Woods, and the macabre things that go on within them, that means they are ever-ready to duck for cover in one of the many hides that they've set up in ditches and hedges about the place.

Clapham Woods came to the attention of the nation's supernaturalists in the late sixties, when a fleet of UFO sightings made the national papers. At the same time, locals started to report strange behaviour from their dogs when they walked through the area where the objects had been seen. Then, a couple of dogs went missing. Straight away, local paranormal groups linked the canine disappearances to the space-ship sightings. People began to chatter excitedly about an other-worldly race of super-beings that had traversed vast universes in order to study life on earth.

But it struck Charles – then a keen 'unexplained' hobbyist and member of Sussex Skywatchers – that if an other-worldly race of super-beings *had* traversed vast universes to study life on earth, it was unlikely that they'd come all this way just to steal two dogs. So, on reflection, he decided to dismiss the UFO sightings altogether. He was convinced that there must be a human explanation for the dog trouble, and set out to find out what it was. He started knocking on doors in the area and, to his rising suspicion, nobody would even discuss it, let alone help. So, at a loss, Charles decided to place his advert in the local paper.

That night, the man in the bush told Charles about a powerful local satanic cult. It was made up of important people – magistrates, doctors, lawyers – who regularly met up in Clapham Woods to perform evil rituals. He said they had powerful political connections, that huge sums of money were involved, and that they were using the forces of darkness to control others. Their idol was a triple-headed goddess called Hecate (pronounced Hek-art-eigh). As well as being Queen of the Witches, Hecate was a notorious dog-enthusiast. One of her heads was a dog, and when she appeared to her disciples during their black masses, it was said that she was followed by a yapping pack of demonic hounds. The neighbourhood pets that were going missing, Charles was told, were being sacrificed by the 'Friends of Hecate', who also had sidelines in blackmail, drugs and sex.

'He told me not to investigate any further,' says Charles as the three of us start walking back up the path to where his car is parked, 'and then he just disappeared.' Charles pauses and gives me a look. 'It was as if he'd never been there.'

But he didn't stop his investigations. It was true, he discovered, that Hecate was a genuine dog goddess. And an associate in Brighton – a doctor – who was an expert in paganism and witchcraft, knew all about her sinister cabal of 'Friends'.

'He said, "For God's sake, don't get involved,"' Charles says. 'He said, "You're dealing with a serious organisation who will not allow themselves to be exposed."'

As we carry on up the path, Charles tells me that shortly after his conversation with the expert doctor he decided to team up with his friend Dave to start patrolling the woods. To this day, the two anti-Satan vigilantes regularly stay up all night to keep watch on the place. And they often find candles, pentagrams, remains of fires, runes, bones and 'tree hexing' – twigs and branches bound together in special ritual arrangements.

Of the two of them, it's shop-worker Dave who appears to be the most prepared. A short, grey, balding man, he's kitted out in combat trousers with a thick brown leather belt, a fawn fleece and a camouflaged hat. Charles, by day a full-time carer, is more casual, in granite-coloured jeans and an old shirt that holds a ten-pack of Superkings and a disposable lighter precariously in its baggy, open pocket. He has long, tangled hair that hangs off his skull in several different directions, leaving his scalp covered in a system of partings that looks like the delta of a desert river. His skin is withered, ashen and loose, and it makes his blue eyes appear so vivid that they look as if they've been popped in by a taxidermist. Both men carry seven-foot wooden staffs that have been lovingly smoothed and would be perfect for caving in the skulls of any devil-worshippers that may decide to launch out of a shrub and attack at any moment. Charles has a crystal embedded in the top of his. 'Yeah,' he tells me when I ask him about it. 'It's been sorted by a wiccan.'

Tonight, they've promised to take me to the most dangerous part of the forest – up the hill to the 'altar tree', which is a hot centre of occult practice. In the past, they've found much satanic paraphernalia there as well as a pool of spilled blood and, once, the decapitated head of a cat. Later on, we're planning to do a Ouija board on a plague pit.

Before all that, however, the anti-Satan vigilantes have to nip to the car to fetch their tea and biscuits. As we come to the end of the rutted and pebble-strewn path, and emerge onto the road, the remarkable tinyness of Clapham village, which bisects the woods halfway up the hill, becomes apparent. The entire place comprises just one street, and the name of that street is The Street. And its residents, according to Charles, are all deeply involved in the bad magic.

'Oh yes,' he tells me, his boots crunching on the gravel as we go. 'And they get hostile, too. Sometimes they actually confront you; other times it's just a general sort of keep-an-eye-on-you type of thing. We've had one incident recently, with a local saying, "Make sure you stick to the footpath, there's going to be shooting up there tonight."' He pauses. 'There wasn't. But it was a warning, a scare tactic, to try and get us to give it up and go home. Well, sorry,' Charles says, jutting his chin in the air, defiantly, 'they've got the wrong people.'

'Are there actual individual villagers that you suspect?'

'Yes,' he says. 'A definite yes. None that I could name to you, though. Proving it, that's part of our job now.'

Just as I'm about to ask my next question, Charles motions me to silence. I glance around for an explanation. Dave's walking stiffly, his small head fixed straight down at his boots. Charles starts whistling, thinly, through his teeth.

'What ... ?' I say.

'Keep ... shush, ssshh,' says Charles.

We're about to walk past a man who's still out working in the warm, failing twilight. He pauses as we pass and leans on the handle of his hoe and his ruddy, earth-bruised face smiles a grotesque run of farmyard teeth in our direction.

'Y'all right?' he says.

'Hello, Bill,' Charles says.

The man nods. 'Going ghosthunting, then?' he says, smiling stilly and staring.

'Oh, no,' says Charles. 'No ghosts round here.'

'Good luck,' says Bill, still smiling.

We continue up the path in silence, until we're out of his eyeshot.

'Phew,' exhales Dave. 'That was close.'

'Was he a Satanist?' I ask.

'We have our suspicions. The trouble is, we're both well known in the village,' Charles explains as we approach his small grey car. 'And every time, every single time, you get: "You going ghosthunting, then?" And it doesn't matter how many times you say no, they still ask you the same bloody question.'

There's a good reason for Charles and Dave's fear of the locals. They believe that people who get too close to the truth about the Satanists are dealt with mercilessly. On Halloween 1978, the local vicar, Reverend Harry Snelling, went missing on his way back from the dentist. A full police search, with sniffer dogs and light aircraft, turned up nothing. His body was eventually found on the Downs in 1981. His death remains unexplained.

'And there were others,' says Charles. 'There was also Nobby Goldsmith – he was the copper. And then there was a woman called Foster. Must've been at least five in all.'

I wonder if the Friends of Hecate are aware of the work of Charles and Dave.

'Oh yes,' he says, with his crystal-topped staff bobbing along, reflecting the stars as it goes. 'Trust me. If you're in their way and they want to stop you, then they will stop you. And they've tried to get at me, I can tell you.'

'What sort of things have they done?' I ask.

He's silent for a couple of steps as we trudge on up the path. 'Things like knocking me off my bike, purposefully and intentionally,' he says. 'And there was another incident. It was in Worthing town centre, in Chapel Road. Somebody stopped me in the middle of the street with a gun. He told me, "Pack it in, or you're going to get this and so is your family."'

'What did you do?' I ask.

Charles lets out a tough-guy laugh. 'Ha! Me, being threatened? I've been threatened before, so it wasn't a surprise. Still,' he pauses for a beat, 'it was a bloody shock that it was a gun.'

I wonder, as we walk, if all this is why he and Dave appear to be a little bit jumpy tonight.

'Jumpy?' says Charles. 'Oh no. There's not time to get scared, is there, Dave?'

'Jesus!' Dave whispers suddenly. 'Shhh!'

A dirty green Land Rover is approaching. We watch in silence as it drives past us and pulls up at the other side of Charles's car, which Dave hides behind.

'Don't say a word,' Charles whispers. 'This could be trouble.'

A large man in Wellington boots and tattered green wax jacket climbs out of the high, mud-splattered vehicle. He stands still and regards us, exuding the rude and bloody confidence of a true man of the countryside.

'Going ghosthunting, then?' he says, with thick, knotty fingers and dry slurry cracking off his boots.

'You might as well get up,' Charles mutters to Dave. 'He's seen you.'

Dave's head emerges from the protection of the car's matronly rear-end.

'Er ... no,' he says. 'No, we're off in a minute.'

We walk past the farmer.

'He's another,' Charles says out of the corner of his mouth.

After they've got their supplies, we make our way up a track towards the twelfth-century church, which has sat alone in the woodland since the original Clapham village that it served was deserted during the plague. All that's left of old Clapham now are rubbled remains amongst the thicket and the pit where the victims of the great sickness were dumped when it swept through the country in the fourteenth century.

The twilight has been completely overwhelmed now, and the ranks of individual trees that stand ahead of us have merged in the darkness into one vast, swaying silhouette. I'm suddenly aware of how small our voices sound. We're breaking a silence that stretches for miles around us and, as we approach them, I feel dwarfed by the army of brooding birches.

'How do we know that there aren't Satanists in the woods right now?' I ask, as the shadow of the forest falls over our faces.

'They could be here now,' he says, 'and unless you heard something or they had a fire going, you'd never know there was anybody there.'

We turn a corner onto a steeper path that, for a few yards, takes us into more open land. And as we do, I get a fleeting whiff of smoke.

'Is that smoke?' I ask.

Dave stops, dead still. Charles sniffs.

'There might be a bit of a fire going on somewhere, yes,' he says. 'That would be interesting, if it was coming from the altar tree. Crikey, that would be just your luck if there's a ritual going on up there.'

'Oh, before we go to the tree, Dave,' says Charles, 'we'll take Will to the stump, see what he makes of that.'

We turn off into an area of sparse, spindly woodland. Between the trees, under our feet, are the crunchy leftovers of several small fires.

'Are these the remains of rituals?' I ask.

'Could be, yes,' Charles says. 'We have found quite a few bits and pieces up here. Look,' he says, stopping suddenly and pulling a flourish of long grass away from a cut tree stump, 'what do you make of this?'

A pentagram has been carved into the wood. And it's not been done idly by some lost local herbet. Its channels are perfectly straight. It is deep, clean and on purpose.

'A five-point star is very difficult to draw, let alone carve,' Charles says. 'This has been done very nicely.' He takes a digital camera out of his pocket and switches it on. 'Oh, you wouldn't flaming believe it, would you?' he says. 'This is always happening. New batteries, they always go down up here.'

Suddenly, I get a more definite wisp of smoke.

Charles stands upright and angles his face up towards the sky. 'Do you know what?' he says. 'I think there is a bit of a fire going on somewhere.'

We all stand there, sniffing the air for clues and, eventually, decide to press on regardless, in the direction of the altar tree.

'People have had stomach cramp, nausea and pressure on the ears in this vicinity,' says Dave, as we go. 'We've had orb pictures, a black shadow, mists. Mediums we've brought up here don't like it one bit.'

'The theory behind it,' says Charles, 'is that they can use the power to affect any living thing. And if the experience of the local dogs is anything to go by, that's true. They can. It works. The only thing I can relate it to is electricity. Now, you can use electricity for good – it gives you light, you do your cooking. But in America, they use it for bad, don't they? They frazzle you.'

'The trouble is,' Dave continues softly, as we trudge up the hill, 'they'll come up here, they'll do their business and they won't shut the energy down that they've built up. So that energy is still hurtling around the vicinity. Anybody out walking their dog could walk straight into that. Responsible occultists would dismiss the forces they've created. But not this lot. They'll build it up and bugger off home.'

'Do you know what?' Charles says, stopping and resting his hands on his hips. 'I think that smoke is coming from the tree. The wind's going in the right direction.'

We stop behind him and look at the smoke curling above the tree-tops and fogging up the stars.

'We might be in trouble,' says Charles.

'What do you mean by "trouble"?' I ask.

'Well, if it is coming from up there, is it something going on? If there has been something going on and there's just remains, that's good. But if they're still up there, that's not quite so good.'

'What would you do if you found them in the middle of a ritual?' I ask.

'Keep an eye on it,' Charles says, and looks at me out of the corner of his eye.

'If they were killing animals ... then we'd probably have to step in.'

'Have you ever found anybody doing anything suspicious?' I ask.

'We've seen people coming out of this area with snooker cases,' he says.

'What do you think was in them?' I ask.

'Ritual regalia, swords, remains of something. Who knows?'

By now, we've reached the corner to the track that leads to the altar tree. We're surrounded by forest, enveloped in the pitch-black night with just the moon glowing above us for company. Dave has crouched down on his knees and is looking up the path towards the fire.

'What do you think?' he says. 'Creep up and have a look?'

'Yeah,' says Charles. He switches his torch on and shines it down at his partner. Dave pulls his camouflaged hat tightly over his head. I can just make out the symbol on the buckle of his brown leather belt. It's a Boy Scouts logo.

'But aren't you worried it might be Friends of Hecate?' I say.

'Well, you don't know, do you?' whispers Charles. 'But we have to have a look. This will be the first time that we've ever done this ... but let's do it. They won't be expecting anybody.'

'We've got nothing to lose,' says Dave.

There's a silence.

'Except our lives,' says Charles.

We creep, silently, up the earthen track. As we get closer, we can make out the large fire through the gaps between the trunks. You can see its fat red belly breathing and its yellow flames groping and spitting sparks up into the canopy.

'If we start hearing moaning or chanting, then we'll know,' Charles whispers.

'Do you really think they'll be chanting?' I say.

'I would've thought so,' he says.

Then, a silhouette of someone appears in front of the fire. It's a man. And behind him, a tent.

'A tent!' says Dave.

'A tent!' says Charles.

'Fuck!' says Dave. 'Dim the light!'

'It's dimmed!' says Charles.

We all freeze and listen, for a moment, to the crackling and murmuring from the area around the altar tree.

'I don't care who it is!' Charles hisses. 'I'm going to go up to that tree and have a cup of tea!'

'We can't do that!' says Dave.

There's a tense silence as we all try to make out each other's expressions in the dark.

'Tell you what,' Charles whispers to Dave. 'Why don't you creep up a bit further and report back?'

'All right,' says Dave, pulling his hat down again, over his face. 'But keep that light dimmed.'

'Okey-pokey,' Charles says, and we watch Dave scrabble up the ditch on his stomach. While Dave's gone, Charles lights up another Superking and starts to tell me why Clapham is such an important location for the local occultists. Over thirty years of exhaustive investigations, he has worked out that there are two main reasons why Friends of Hecate chose Clapham Woods as their base.

The first is that six powerful ley-lines intersect them. This rare and numerically significant conflux (six being the 'number of the beast') creates a 'black stream' which is ideal, he believes, for devil-worshipping and malevolent conjuration. The second reason is that there's a lay-by on the nearby A27 that the Satanists find handy for parking.

I watch Dave creep up the track and, as the dangerous dark folds in around him, it strikes me that everybody needs a Satanist. Because knowing whose fault it is can be a great comfort. It helps define you, knowing who you're not. It's reassuring. Especially if you're an anti-Satan vigilante or, indeed, a Christian. And then it occurs to me that enemies can also be exciting. The thrill of having the spicy breath of the dark side prickling the skin on the back of your neck can be seductive and thrilling and vital. Isn't that one of the reasons why we want to believe in the supernatural, in the devil and in evil and in ghosts? Because their very presence in our days makes our lives feel less ordinary? And the fact that they never quite get you has the perverse effect of making you feel safer. Is this just another chimera, invented by the brain, to make us zombies feel alive and to stop us blowing our brains out? It's hard to deny that these diabolical enemies are an excellent ointment for the self-esteem. After all, if something that powerful and frightening and unconquerable can be after *you* ... well, you must be pretty important. Not inconsequential or lonely. Not inadequate or small. You're obviously not just a zombie, living an automatic life. There's just no way that you're here for no reason at all. If for nothing else, you're here to do battle. It's a war, an epic fight against evil. And that makes you the hero.

'I can see that fag butt from right up there!'

It's Dave. He's arrived back, on his hands and knees, in the ditch.

'What are they up to?' Charles whispers, shining his torch down on his friend.

'Milling,' Dave says. There's a silence. Dave adjusts his hat. There's a small twig stuck to his left cheek.

'Hmm, milling.' Charles says. 'No chanting?' he says. 'No moaning?'

'Not that I could hear.'

'But could you actually tell what they were up to, Dave?'

Dave looks at Charles.

'They might just be camping,' he says. 'Kids ... camping'.

'Well, bollocks to them,' he says. 'I'm going to just march right up to them. I don't care.'

'I don't see how we can,' Dave says.

In the distance, the sound of teenage laughing spirits through the trees. We listen to it in silence. Eventually, Charles and Dave decide that perhaps it's best if we leave the altar tree after all, and we trudge off to do our Ouija board.

I can't pretend I'm not a little apprehensive about this. Almost everyone I've met on my search so far has warned me against the Ouija board. They've said it's an unfinished device, that you can't control it, that you risk possession, madness or suicidal depression if you even dare tinker. Priests, parents, druids – even the lucid and rational Lance from the Ghost Club – have all told me to keep well away from the crafty seductions of evil Ouija.

Nobody knows who invented it, but the talking board became mainstream during the Spiritualism movement in the mid-nineteenth century. In 1890, two businessmen, Elijah Bond and Charles Kennard, had the idea of patenting a set and an employee of theirs, William Fuld, came up with the term 'Ouija' (which comes from the French and German words for 'yes'). In 1901, Fuld took over production of the boards, and in 1966, Fuld's estate sold the patent to toy-makers Parker Brothers, who own the trademark on the name to this day. Nobody disagrees that the planchette on a Ouija board does appear to move without any of the participants being consciously responsible. Sceptics, however, theorise that players are making tiny, involuntary muscle movements (the so-called 'ideomotor effect'). Others think that spirits are using the energy of the participants to communicate and, sometimes, materialise. Christians believe it's the devil.

After a short walk, we find ourselves crouched low on the ground in an abandoned, pathless acre of the ancient Sussex woodland. It's testament to how close-knit the community is round here that they know the location of this plague pit, one of the many forgotten mass-graves that silently haunt Britain's subterranean landscape. All that's visible now is a barely noticeable, twenty-foot-square drop. I watch Charles brush some dead, coppery undergrowth from the board, as the birches huddle and whisper above us. Charles picked up this 'Waddingtons Mystifying Oracle' in a charity shop that he used to work in. He claimed it for himself when they refused to sell it. Right now, it's illuminated only by a small candle, whose thin, cowering light is being diluted by the shadows and bullied by the breeze.

'We ready, then?' says Charles as the flame sends its weak yellow light over his long hair and watery eyes.

'Yep,' I say, and we each place a fingertip on an upturned shot-glass in the middle of the board.

'Right then,' says Charles. Almost instantly, the glass shoots across the board and starts to spell out its message. I try to follow it with frantic jerks of my head.

C to L to A to P to H to A to M.

'Clapham,' says Charles.

E to V to I to L

'Evil,' he says. 'Clapham evil. Hmmm.'

The glass feels as if it's floating on a skin of oil. And there's a real force coming from it.

'What do you mean by that?' Dave asks the board.

It moves again. Y to O to U to D to A to N to G to E to R.

'What from?' he says.

P to E to O to P to L to E.

'Well, yeah,' says Charles. 'People. What type of people? Who? Who?'

D to E to V to I to L to W to O to R to S to H to I to P to P to E to R to S .

'Devil worshippers ... ' says Charles.

T to H to E to Y to A to R to E to A to W to A to R to E.

'Jesus!' says Dave.

T to H to A to T to Y to O to U to A to R to E to H to E to R to E.

'Interesting,' says Charles. 'They're aware that we're here. Hmm. They must have seen us come up. Or they're in the woods somewhere. Who are we speaking to, please?'

R to E to V to H to A to R to R to Y.

'Is this the Reverend Harry Snelling?' says Charles.

The glass shoots to 'Yes'.

I to A to M to S to T to I to L to L to H to E to R to E.

'Who's keeping you here?' I say.

F to R to I to E to N to D to S to O to F to H to E to C to A to T to E.

'Why are they keeping you here?' Charles says.

As the dead reverend tells us that THEY ARE USING ME FOR MY ENERGY, I notice that there's a crescent of white on the end of Charles's fingertip, where mine and Dave's are pink.

'So has everyone got their fingers on the glass quite lightly?' I ask.

'All lightly, yes,' Charles says.

'It doesn't feel heavy, does it?' I say.

'It's flying,' Dave says. 'I've never seen it as strong as this. Usually it takes a while to get going. But this is very strong. This just went straight into it.'

My phone starts buzzing in my pocket. Now I'm really scared. It's too late for this to be anything but bad news. I reach into my jacket with my spare hand and pull it out. On the screen is the word 'Home'. Farrah.

'Sorry,' I say, 'I should take this.'

I put the phone to my ear.

'What's up, love?' I say. 'I'm in the middle of something.'

'You know the stereo in the kitchen?' she says. 'It keeps turning itself on and off.'

'What?'

'I keep turning it off and as soon as I leave the room I hear it turning itself on again. It keeps happening.'

'When's this been happening?' I say, with my right index finger still outstretched on the Ouija board. 'Just now?'

'Yes,' she says. 'The last few minutes. Will, I'm shitting myself. I keep thinking you've brought one of your ghosts home … '

'Just turn it off at the plug,' I say, 'and call me back if anything else happens.'

I finish the call, put the phone back in my pocket and smile flatly at Charles and Dave.

Next, I decide to test the spirit's knowledge. If this really is a citizen of the invisible world, then it will know this.

'What's my girlfriend's name?' I ask.

S to U to S to A to N.

'It's not Susan,' I say.

There's a silence.

I to S to H to E to R to E.

'Susan is here,' says Charles.

'Who's Susan?' I ask.

'I've no idea,' Charles says. 'Is anybody else there?'

The glass shoots to 'Yes', then D to O to R to E to E to N.

'Doreen?' I say.

'There's only one Doreen I know,' says Charles.

Dave looks up at Charles. 'It can't be,' he says. 'It can't be Doreen Valiente.'

'Who's Doreen Valiente?' I say.

'She was a local witch,' Dave says.

'Well ... local?' Charles says, with a humph. 'International.'

Doreen the International Witch starts telling us that we're in danger, and as the glass moves away from Charles, it tips suddenly, at the end of his finger. Then, shortly afterwards, when the glass is moving towards him, it pulls over again in his direction. The tip of his finger, I notice, is bloodless throughout.

Suspicion begins to nibble at me, like a goldfish going at breadcrumbs. Could Charles, I think, be pushing the glass?

'Can this work with just two people?' I ask.

'Yes,' says Charles.

'Shall we try and make it work, then,' I say, 'just with me and Dave?'

'All right,' says Charles. He takes his finger off and sits back and looks at us with his arms crossed.

'Are we in danger?' I ask.

Nothing happens. There's a silence.

'Come back to us,' Dave says.

Nothing happens again.

'We've had this before,' Charles says. 'When Dave goes on with people, it doesn't work.'

Dave looks at me mournfully as nothing carries on happening. 'It's almost as if I'm blocking it,' he says.

We sit there in silence for a moment. Charles watches us from the shadows as our fingers stretch out in front of us, uselessly. A breeze lifts through the trees and their leaves start rustling a mocking applause. I begin to feel a bit sorry for Charles. If he *was* pushing it, maybe it was just because he was desperate for it to work in front of me and didn't want to risk it by relying on unreliable spirits. There again, what if I'm being unfair? Maybe whatever power was making it move *has* just ceased, and if Charles rejoins us, it won't make any difference – the board will stay lifeless.

'Maybe you should come back on, Charles,' I say.

And the instant that he does, the glass shoots back into action.

'Hello, Janet?'

'Hello.'

'It's Will here. Do you remember, we had an arrangement to have an interview?'

'Oh, yes.'

'Well, it's been a couple of months now, and I was wondering ... '

'Sorry. I haven't forgotten. I will be in touch.'

'Do you have any idea when?'

'Not at the moment, no.'

'Oh, OK. Thanks, Janet. Goodbye.'

'Goodbye.'

13

'We've got strangers in the house'

In a dismal Wetherspoons in a town ten miles north of London, Milton slips gently into a back corner booth. He pours his tonic water carefully out of a little bottle, which has deep scratches, from use and reuse, calloused around it. The liquid fizzes into a thin, cheap glass that has three or four melting ice cubes, which look like half-sucked Glacier Mints plopped sloppily into the bottom. Once he's finished pouring, with the tiniest facial movement, he registers surprise at how little of the bitter effervescent water there is in the glass.

'Do you want me to get you another?' I say.

'No,' he says. 'That's OK.'

He's got to leave soon. It's a golden Sunday morning outside of this alcoholics' grotto, and he's anxious to get out of here, back to his wife and son. I turn my tape recorder on and he looks at it. An involuntary, embarrassed smile blooms across his face.

Suddenly, I feel awkward. In that instant, I become overpoweringly aware that Milton didn't want his story on the record, and that I've sort of nagged him into it. He takes a sip of his tonic water, partly, I suspect, to pull the attention away from the moment.

'OK,' I ask, 'how old were you when you saw it?'

'I can't be certain,' he says, putting his drink down and his hands on the table, 'but I reckon I must have been around five or six years old.'

'And had you just moved into the house?'

'No,' he says, 'we lived in this house for quite a while.'

'And this house was in north London?'

'It was a house in Birmingham, Edgbaston.'

'Was it a modern house?'

'No, it was an old terrace house.'

'And you were in the bath?'

'No, I wasn't in the bath. I was in the bathroom.'

There's a small pause. Milton glances, again, at the red light on my tape recorder.

'Hmmm,' I say. 'I'll tell you what. Why don't you just tell me the story in your own words?'

'OK,' he says, and he takes another sip and begins.

'I'm sure it must have been a weekend, because all the family were in. It was a bright, sunny day. It was really hot and my mum was outside, hanging the washing on the line, and my dad was outside, too. I don't know where my sister was. Me and my brother, who was two years older than me – we were playing in the house. We went upstairs, into the bathroom, and we started mucking about. We were getting bits of toilet paper, putting it under the tap, and making balls and flicking it at each other. We were, like, howling with laughter, we were really happy and elated, when all of a sudden, we saw something happening by the window, so we stopped. We both looked, and the way I recall it is that two figures were, like, materialising. I can't recall the actual materialisation. My brother's talked about this before, and he reckons that they, like, floated upwards, and appeared. In my mind, I see them being faint and becoming more clearer. At the end of the materialisation, there were two ladies standing in front of us. They were solid, you couldn't see through them, they were like normal people. Their ages – they weren't like old ladies. They weren't young. They were middle-aged. The clothes I remember them wearing, I'd say it was like 1930s, 1940s kind of clothing. And I think that they had hats on, and they were smiling at each other and looking at us and they were really happy. We were just stood there. We knew it was impossible. Things like this don't happen every day. But they were there and we didn't feel they were going to harm us or anything, they were just smiling at us, staring at us. And I said, "Do you want us to get our mummy and daddy?" because, as far as I was concerned, they were adults, and when there's an adult in the house, you've got to get your mum and dad and tell them. So they said, "Oh yes, go and get your mum and dad."

'So we both ran out. Ran down the stairs, ran outside. "Mum! Mum! Dad! Dad! We've got strangers in the house." And my mum just carried on putting the washing on the line and said, "Go and tell your dad." So we ran and told our dad and he was sawing wood. I remember it clearly because of the smell of the wood, and it was a bright, blistering hot day, and I was quite shocked that no one would believe us. It was like, hang on, there's people in the house, we don't know who they are, they could be walking around, going in our things. There's people in the house, we're telling you, and you're not believing us. We just couldn't believe it. So we ran back into the house, as fast as we could, back into the bathroom and we looked and they were gone.'

And there it is. The best ghost story I've ever heard. It's great because it appears to defy all the psychiatric theories I've been told. Although they were clearly in a state of high arousal, Milton and his brother were not expecting to see ghosts. They weren't in a spooky location where the *feng shui* was supernaturally significant. And most of all, they're definitely not the kind of people who are 'the type' to believe in the paranormal. I've been on a vigil with Milton, and he is the most sceptical investigator I've ever seen. He was unwavering in his unimpressedness, dogmatic in his doubting. Really, on every level, this is not your typical ghost enthusiast: he's not defensive or boastful, he doesn't wear supernaturally significant pendants, crystals or rings, doesn't have a Messiah complex or voices in his head, he's got normal hair. In fact, Milton is only a member of the Ghost Club at all because he wants to find an explanation for what happened on that strange summer morning. And, throughout all the vigils he's ever been on, he's never seen another ghost. He did, in a pub in East Sussex one night, see a door handle move by itself (the door was open, against a wall, and he was on the other side of it). But that's it, in the thirty years since the spectral women.

I'm troubled by Milton's tale. It's pulling me back in. You see, I can't explain his experience. I can't find a way around his story. Every sceptical path through it is blocked by another detail. It's not one of Jung's hypnogogic states because he was nowhere near sleep. It's not a hallucination, because his brother saw the same thing. It's not a 'Stone Tape' incident, because the women interacted with the brothers. It's not a result of mental illness or any sort of personality defect, because it's not part of a pattern of behaviour. And it's not a false memory, because his brother remembers the same thing.

'He feels uncomfortable talking about it,' says Milton, before tipping his head back and milking the last drops from his glass. Outside the pub, the streets have suddenly darkened. The first sopping bullets of a flash rain-storm start hitting the warm street. 'I talked to him about it through the email. I told him I was going to do this interview with you and I asked him if he minded, and said that we'd keep his name out of it. He knows it happened. Whenever we do talk about it, he goes, "Yeah, yeah, it happened," but he can't explain it. We're both logical people, but I know that illogical things happen in this world. He doesn't really want to accept that. If he can't answer it, then he doesn't want to think about it. If you don't talk about it, it's out of your mind, isn't it?'

TO: admin@paranormaldimensions.com
FROM: Will Storr
SUBJECT: Stan

Hello

I wonder if you can help me. I saw a report in your local newspaper about Stan and the ghosts he was seeing on a nightly basis. I understand that you are a sceptical investigation group and you came to the conclusion that Stan's ghosts were something to do with a hypnogogic state. I was wondering if there would be any chance of meeting you and Stan to discuss the case for some research that I am doing?

Many thanks.
Will Storr

'Hello Stan, my name's Will and I've been working on some research about ghosts.'

'Oh, right.'

'I read about your problem in your local paper.'

'Oh, yes?'

'And I was wondering if I could come up and interview you about it.'

'I suppose so. That should be fine.'

'I'm coming up to see Paranormal Dimensions as well.'

'Paranormal Dim … ? Oh yes. To be honest, that lot, they talked a load of rubbish.'

'Did they not solve the problem, then?'

'No, I called the local vicar round and he sorted it out. He did an exorcism. It's been fine ever since.'

'Oh, OK. Could I possibly have the phone number of the vicar?'

14

'They called me Ghost Girl'

She wears a thin gold necklace with a pendant that rests in the cup of her collarbone. It spells out her name in happy, bulbous cartoon script. The letters increase in size, from the first to the last, and they glint in the light as the movement of her slow talking makes them shift on the surface of her skin. Its comic rendering is like the logo of an iconic superhero and, over the months, that's almost what this pretty, firm and frequently sorrowful woman has become to me.

Janet. I've finally been granted an audience with Janet.

As I sit here, on a creaking wicker chair, between two cat baskets and a washing machine, it feels as though this boxy conservatory in Clacton-on-Sea is the destination of a long pilgrimage. Janet holds enormous power because she holds all the answers. The endless battle between the sceptics and the believers raged around her when she was eleven years old, and she knows who was right and who was wrong. All the answers are here, in front of me, drinking tea and wearing dun-coloured three-quarter-length cargo pants and a tight Nike T-shirt.

'I mean, we're talking about twenty-five years or more, aren't we?' Janet says quietly, with her tough, thin fingers pushed through the handle of her mug. 'I can't remember everything. But I can remember the main events because, you know, they leave scars.'

Janet has a way of looking at me that I find acutely unsettling. It's as if she's run a cold hose straight into my marrow. She keeps reminding me of someone, and I can't think who. It's her fragile, birdlike quality and her gentle, sad way of speaking, but, most of all, it's the cavernous gaze that you'll suddenly find yourself trapped inside. I'm sure I've seen it before.

After my meeting with Maurice, I ended up deciding that the Enfield case was so extreme in what it required me to believe that my credulity just shut down, and I went to see a psychiatrist. But, since that encounter with Dr Salter, there have been a couple of incidents which have caused me to call those uncompromising sceptical beliefs into question. Firstly, there's Milton's testimony. And, secondly, I met up with one of the doctor's recent patients. Frances, the logic goes, is just like Debbie, Derek and all the other mediums and supernaturalists that I've met. The only difference is that Frances has been cured. My meeting with her was sad and surprising and terrifying. Frances was a highly regarded teacher when she started hearing voices in her head. And now, months later, she's barely recovered. When I spoke to her, she appeared bashed about with shame and shock over what had happened. It was as if her mental illness was a deranged clown that had jumped into her driving seat and joy-ridden all over her preciously tendered life, with her strapped, a helpless witness, into the backseat. And now he's gone, and all she can do is look over the wreckage and wonder.

Frances would be 'possessed' every day. And, like my paranormalist friends, she'd hear the voices of the things that were possessing her. But Frances's voices told her that she was God and that she'd started the earth. They also told her that an organisation called Atlantis Revisited had been listening to her thoughts and that witches were attacking her, stealing parts of her body. She'd call Scotland Yard obsessively and ask to be put through to the 'Dewitching Department'. And that was just the start of it.

I think there are some important differences between Frances's mental state and those of my witnesses. For one thing, when Frances was going through all this, she was catastrophically paranoid. For another, she was terrified. And for yet another, she was completely out of control. But most of all, Frances was acting without any logic at all. Her mind was loose and free-styling, like an experimental jazz trumpeter lost in space. Debbie, Lou, Dave and all the rest are functioning people. They have beliefs that fit into a pre-existing world with rules. Frances's world, though, was self-created and utterly anarchic. I thought, for a while, that it was simple, that Frances was just *a lot more* unwell than my supernaturalist friends. But then I found out that Frances didn't actually see any witches or ghosts or demons. She didn't even see anything move across a sideboard. So, as profoundly sick as Frances was, she actually experienced far *less* than these other people, who, while doing what they do, continue to be attentive parents, successful small businessmen and lead guitarists in Spider Monkey.

So, my meeting with Janet has come at an expedient time, because if I'm to let the Enfield case make me a true believer, I need to examine it more closely. And the research I've done since my visit to Maurice's has led me to believe there is plenty more examining to be done.

'I remember when it first started,' Janet says, in a voice that's so quiet it's temporarily drowned by the miaow of an overweight ginger cat that's just crawled in through a window behind her. 'I was due to start senior school, I think, the following week. It was the thirty-first, no ... ' she says, and touches her forehead with the tips of her fingers, 'the twenty-eighth of August. No ... it was the thirty-first. The thirty-first of August nineteen seventy-seven. I slept in the back room with my brother Johnny, and we was laying in bed and we sort of heard something and Johnny said, "What's that?"'

Depending on who you believe, this moment marked either the beginning of the most widely witnessed and documented haunting in history, or one of the most successful wind-ups. The night before this, Janet and Johnny had had another small disturbance which their mother had given short shrift. And when they complained this time, she told them, yet again, to get to sleep.

Then the children saw a chest of drawers slide from the side of Janet's bed towards the door.

'We shouted, "Mum! Mum!"' Janet says. 'Because we were sort of frightened, but also intrigued by what we were seeing. It's so vivid, some of it,' she says, as the late autumn sun begins to set over the kids' bikes and plastic tractors in the garden behind her. 'And Mum did see it moving in the end.'

'What was her reaction?' I say.

'Well, she was just dumbfounded, really. And she pushed it back. And then it started to move again – like it was trying to block her out of the room, like it wanted me and Johnny to itself. And she tried to push it back again, and it wouldn't move. So she said, "Right, all get out of your beds and we'll go downstairs." We was very nervy. There was a funny atmosphere in the house. And then the knocking started. And Mum said, "Well, what'll we do now?"'

She decided to fetch her neighbours, Peggy and Vic Nottingham.

'I think they thought there might be someone like a burglar in the house,' Janet says, 'but they couldn't find no one. There was just, like, this knocking. It sounded like it was coming from the outside wall, but it was like it was inside as well. And sometimes it sounded like it was coming from underneath the floorboards.' She puts her mug on a small table by the side of her chair. 'It was a big shock to us all, really. We didn't know how life was going to change.'

Peggy decided to call the police. One of the officers who responded saw a chair rise up and move across the floor on its own, in full view of everyone. 'Well, she was astounded,' Janet says, 'we were all astounded.' But she didn't know how to help. So, at a loss, they decided to call the *Daily Mirror*. 'They stayed for quite a while,' Janet says, 'and nothing much happened. And as soon as they went to go, it all started.'

'What did?' I ask.

'Lego bricks, marbles flying about. And then the photographer from the *Mirror* came back and I remember that very well,' she says, 'because a Lego brick hit him right above the eye. He still had the mark a few days later. And then Maurice Grosse came in on the case. And then there was so many people that come. I think it got to the stage where more and more people were coming in and out, and my mum never had a man behind her because she was divorced and it was starting to become, like … '

Through the window, behind her, one of Janet's children has come into view. He's holding a thin sheet of metal and is standing with his legs apart in front of a life-sized pottery leopard. To occupy himself while he's waiting for his dinner of oven chips and frozen pizza, he wobbles the sheet backwards and forwards so that it makes a sound like Aboriginal thunder music. We're shut away, out here in the conservatory, because Janet doesn't like talking about the haunting in front of the kids. In fact, Janet and her husband have, so far, only told the two older ones about it.

'I mean, you know,' she continues, pushing a line of blonde, waist-length hair out of her face with a finger that has several gold rings around it, 'everybody wanted to come and see it, you know, but that was making things worse for Mum, not being able to cope. She was glad of Maurice's support. But it has been said, right, that he was keeping it going. But, I mean, how can you keep something going like that? It starts and it stops when it wants to, you know?'

It was Anita Gregory who published the theory that Maurice Grosse was somehow 'keeping it going'. She was Enfield's most vigorous critic and, over the last few months, I've been desperately searching for her banned thesis. It's been difficult. The SPR refused to tell me when it was written or which college Anita was attending when she wrote it, except to say that it was possibly a London one. They wouldn't even give me its title. So, I called all the colleges in London. Some of them looked and couldn't find anything; others refused to do even that, because I wasn't a student. Next, I went back to the British Library, as I was told they keep all theses produced in the UK. This, I thought, was my last chance. I sat down at a monitor, logged on to their 'inclusive catalogue' and typed in Gregory, Anita. A book about a medium popped up, which was published in the mid-seventies. And that was it. I just stared at the screen for ages. Then I clicked furiously on the name, over and over, clicking on the mouse. This must be my Anita, I thought. It must be. But where was her thesis? Where was her fucking thesis? I was about to give up when a staff member in remarkable glasses walked past.

'Are all Ph.D. theses stored on here?' I asked.

'Most of them,' he said, touching his thick circular frames, self-consciously, 'but not all of them.' He pointed to a small sub-room in the corner. 'You could try one of the computers in there.'

I'd already given up. But I went in anyway and sat down and logged on again, with the sticky fog of an angry sulk fast descending.

'Gregory,' I typed in. 'A.'

Hundreds of titles fell down the screen in old boxy type.

A. Gregory: 'Farm Income Inequality & Stability'.

I kept scrolling, my eyes running off the titles like rain down a gutter. No, no, no, no. That's not it, no. The first hundred went. Then the second hundred.

A. Gregory: 'The Effects of Barbiturate & Other Sedatives on Fish Retinal Neurones'.

And then I saw it.

A. Gregory: 'Problems In Investigating Psychokinesis In Special Subjects'.

This had to be it. Psychokinesis is the ability to move objects with the power of your mind and if Anita was sceptical about the case, this could well be what she thought was behind it. And 'special subjects' – that could be Janet. And if she was writing about 'problems in investigating' the case, she could well have used Maurice's attempts as examples and that, right there, is why he took such offence at it. I filled in my request form and paced over to the counter to hand it in. The man looked up at me with coal-dead eyes and said that there was no 'shelf number' on the entry. And that meant they didn't have it. But … but … but … it's really important, I said. He replied that he was sorry. But he didn't look sorry. And then I went home, in the rain.

So, I decided to join the SPR. Perhaps, I thought, there might be something in their archives that could give me a clue as to what Anita had said that enraged Maurice so much. Eventually, I discovered a series of letters to and from Grosse and Gregory. Dated from the early eighties, they were published over several editions of the SPR's quarterly journal. I was thrilled. This was the argument that must have led to Maurice's patience finally erupting. I read the letters with a mixture of triumph, fear and fascination all boiling up in my blood. By the time I'd finished, that toxic slurry of emotions had fermented into pure liquid shock. It was devastating. The longer the polite but furious scrap went on, the more sceptical Anita became. And by the end, she had become extremely sceptical. Over the course of the exchange, she says, in print, in front of all of Maurice's SPR colleagues, among them the most esteemed paranormally curious minds in the world, that the evidence gathered at Enfield is 'questionable', 'greatly exaggerated' and, ultimately, 'pathetic'.

In one of the letters, Anita claims that she interviewed WPC Carolyn Heeps, the police officer who saw the chair move. Apparently, she told Gregory that she thought it was the children playing tricks. Anita also records a comment made to her by the neighbour, Peggy Nottingham, on 15 January 1978. 'Mrs Nottingham told me that what was going on now was "pure nonsense" and it was "kept going by the investigators".'

I ask Janet, 'Do you remember Anita Gregory?'

'I think I can remember her having black hair and glasses,' she says. 'She was very much a sceptic. I think she found the voices … I mean, she seemed OK, but it was like, what was going on was beyond her – it couldn't possibly happen. And she went away very, very sceptic.'

I can understand why. I pull photocopies of the correspondence out of my bag and read Janet a section where Anita comments on the fact that nobody was allowed in the same room as her and her sister when they were speaking in the gruff 'old man's voice' that was supposedly possessing them. I also find this incident dodgy in the extreme.

'Yes,' Janet says, 'I think Anita Gregory thought that as well. But in the beginning, it was pretty much the same thing. Like, the lights had to be out before the knocking started.'

I look at Janet and think about what she's just said. I still find it dodgy. Not least because there's worse to come. You see, Anita was eventually allowed in the bedroom to ask questions. 'Provided, that is,' she wrote, 'I faced the door and covered my head with the girls' dressing gowns.' And when she did, 'slippers and pillows were shied at me'.

Janet smiles and then breaks out into a tender laugh. It's as if she's amused by her younger self and greatly affectionate towards her.

'I can't remember that,' she says.

'But you do remember the fact that people had to be looking away before the voice spoke?' I ask.

'I think that maybe it didn't want her to see because she would go away sceptic.'

'But surely if it let her watch, then there would be *more* chance of her believing it?' I say.

Janet's eyes widen. 'But Graham Morris had the proof,' she says, 'he had the photos.'

'I know there are photos,' I say, 'but I'm still not clear why you wanted people out of the room before the "voice" spoke. Was it you talking when they were told to leave? Or was it the voice of the spirit, talking through you?'

'The voice of the spirit,' she says.

'So you don't know why it asked people to leave?' I suggest.

Shit.

'No,' Janet says.

I'm an idiot. I've just fed her the 'right' answer, the one I thought would explain it best. Inwardly, I'm furious with myself.

I stare, angrily, at the ginger cat on the floor and decide to move on, to the most damning evidence that Anita provides. This concerns an incident where the researcher David Robertson set up some hidden video equipment and filmed Janet bending a spoon with her hands and then jumping up and down on her bed and flapping her arms like wings. She was, Anita says, caught 'merrily cheating away and giving not the slightest indication that she was aware of being filmed'.

'I remember that one,' Janet says, smiling, blushing and nodding deeply. 'I thought, Christ, here we go. Maurice *was* annoyed with me. Yes, I remember that well. I had some felt-tip pens in my hand, he come down and he says something like, "I'm not very pleased with you, Janet."'

'So why did you do fake stuff?' I ask.

'Well,' she says, her hands cupped on her lap in front of her, 'there was times when things would happen and times when they wouldn't and sometimes, if things didn't happen, you'd feel that somehow you'd failed.'

'So you'd feel obliged to make things up?' I ask. 'Because you didn't want people to be sceptical?'

'Yeah,' she says, 'that's pretty much it. Plus you'd get bored and you'd get frustrated at all the people coming and going. I mean, life wasn't normal. But the incident you were saying about, David Robertson, even at that age I thought, God, what's he on about? You have more respect for people as you grow up, you know, but there's some people you just can't stand.'

'So, you decided to wind him up because you didn't like him?'

'Yeah. He was just sort of ... dozy.'

So eleven-year-old Janet, already frustrated at the disruption to her life, comes across yet another researcher creeping about in her house and implying she's a liar. And this one really irritates her. So she decides to fuck with him. This, I think, I can believe. It's just so human. Moreover, it's just so eleven-year-old human.

'How much of the phenomena at Enfield was faked, then?' I say. 'As a percentage?'

'Hmm,' Janet exhales and looks into her lap. Her chin creases with the concentration. 'I'd say two per cent.'

I decide to hit her with the rest of Anita's moody opinion, just to see what'll happen. 'It wasn't just the voices, you know. Did you know that Anita ended up being sceptical about all of it?' I ask her, flicking through the sheets on my lap and reading out comments that I've underlined. 'She says it "withers away under close inspection", that it was a "poorly researched and doubtful case", that it was "pathetic" … '

By the time I look up, lines of hurt and anger have tightened Janet's face.

'Yeah, but she wasn't the one there, was she?' she says. 'She wasn't the one who experienced it. There's so many people that did come round and see things. What's she saying, then? They're all pathetic? They're all liars? They're all wrong?'

Certainly, I think there was nothing mentally wrong with Janet at the time: the neuro-psychiatrist Dr Peter Fenwick and his team submitted her to a battery of tests at the prestigious Maudsley Hospital and confirmed that there was nothing broken in her brain.

It's almost ten o'clock now, and the last train from Clacton leaves in half an hour. I don't want to get stranded here, amongst the cats and the kids and their mother's sad memories. But there is one more thing I want to check before I go. If Lou Gentile was researching this case, it would have been his first question.

'Before this all happened,' I ask, 'did you ever use a Ouija board?'

'Oh, we both did, yeah,' Janet says. 'Me and Rose did it with a couple of friends in the school hut. And when we were doing it, the glass tipped over and smashed and there was, like, a face at the window. We was frightened at the time, but we didn't think anything more of it.'

'Do you think this could have had anything to do with the poltergeist?' I ask.

'It could have done,' she says, with a loose shrug, 'but there can be a lot of coincidences in life, can't there? Mind you, the same thing did happen with my friend and it's only just died down now.'

'What?' I ask. 'Tell me.'

'She did a Ouija board and got the spirit of a young girl. After that, she'd be frightened to stay in the house on her own. She'd hear footsteps and see apparitions. She's got four children and it got to the stage where they didn't want to sleep in their room, you know?'

Janet tells me that her friend eventually got her house straightened out by an exorcism. It was also, she says, a priest's visit that led to the Enfield haunting 'quietening down'. I note to myself that both these cases are bookended with a Ouija board and an exorcism, just as Lou and Father Bill and the rest would predict. Then I look up and just listen to Janet's quiet talking.

'I was always the strong one,' she's saying. 'You look at photos that were taken at the time and I was the one who always smiled through. And yet life wasn't all that. I was bullied at school. They called me Ghost Girl and put crane flies down my back. And I'd dread going home. The front door would be open, there'd be people in and out, you didn't know what to expect or what was happening inside, and I used to worry a lot about Mum and what effect it was having on her. She had a nervous breakdown, in the end. But I'm not one for living in the past. I want to move on. But it does come to me now and again. I dream about it, and then it affects me. I think why, why, why did it happen to us?'

As I listen to Janet, it strikes me that it really doesn't sound like she revelled in the experience, as a hoaxer might. Right now, she sounds like a woman talking about any bitter trauma. In fact, she sounds just like she did when she told me about the death of her son. And, it seems, Janet and the rest of her family have continued to suffer since then. She tells me about her brother, who stayed in Enfield until Mrs Hodgson's death.

'He'd hardly ever go out because he'd have to keep looking

over his shoulder,' Janet says. 'People didn't forget. They'd take the mickey out of him, spit at him. And on the bus, you'd get people peering at you. They wouldn't say nothing, but they'd peer at you. My brother's sensitive to all that. It brought him down a lot. He couldn't go to pubs because he'd get in fights. They'd be like, here he comes, freak boy from the ghost house. He went to college and they'd start on him. In the end, he left. They're more refined here, out of London. People don't know the past.'

Up until her death, Janet's mother would tell her that there was still something in the house. She'd hear footsteps on the stairs, and doors would open and close on their own. She didn't mind so much, though. Compared to the hell that had ripped through the place during those months at the end of the seventies it was easily ignorable.

Janet says, 'Even my brother, until the day he left that place after Mum died, he'd say, "There's still something there." And there was. Put it this way, you'd feel like you were being watched.'

'And do you ever feel there's something still with you?' I ask.

'Sometimes,' she says. She's speaking very softly, reluctantly, now. 'I do. Nothing nasty, but I do. I can honestly say, though, that I'm not possessed. I ... '

She looks at me, and then, with the wrecking ball of realisation almost knocking me through the window, I finally get who it is that Janet reminds me of.

' ... I try to convince myself that I'm not, anyway.'

It's Kathy Ganiel.

'Hello, I wonder if you could help me? My name's Will and I'm doing some research about ghosts.'

'Oh, right.'

'I was just talking to someone you know – Stan. You did an exorcism round his house the other day.'

'Right.'

'And I was just wondering if I could come up and interview you about it.'

'No. That won't be possible.'

'Stan said it would be fine. He doesn't mind –'

'Well, Stan may well have said that, but I'm a professional –'

'But –'

'I'm sorry. Goodbye.'

'Hello, is that Will?'

'Yes. Hello.'

'Hello. It's Stan here.'

'Oh, hi. I've just spoken to your vicar.'

'Yes, actually I just phoned up about that. To cancel our interview.'

'Oh. Oh. Why?'

'Sorry about that. Goodbye.'

'But –'

'Goodbye.'

15

'That's Annie's room'

That view. It would be hers now – hers to look at, whenever. She paused for a moment and looked out at it again, through the small window, and rubbed her hands together for warmth. She liked it down there, way out west. They were nearer to the edge of things. There were right at the edge of the land, for a start, where the ground drops into the ocean and the spray leaps up at it. They were nearer to history, too. For centuries, the only things that had changed around there were the cars and the Christian names. And best of all, they were nearer to nature. The weather in that part of the world was quick and changeable, but on that particular May morning, the sky and the sea had decided to lie back after the furies they'd whipped themselves into during the winter. The sun glittered on top of the Atlantic and slipped easily through the small, thin window in the top room of the First And Last Inn. It looked out towards Sennen Bay and in towards a white-walled double bedroom filled with cardboard boxes and dust that floated like plankton in the shafts of light.

Lynne blew warm breath into her cupped hands. That morning, despite the heat of the day outside, she was freezing. Really, unaccountably cold. Even though she'd pulled on a thick jumper and then a lambswool cardigan on top of that to keep the leaking chill off, she could still feel it. It bled its way under the layers and pushed itself into her skin. Just as she became aware that her fingertips were slipping automatically into her trouser pockets, she stopped herself and decided to go downstairs to see how Neil was getting along on their first day. She picked her way through the boxes that she was in the middle of unpacking, and stepped out of the wooden door, which was stiff and misshapen with age. Moments after she entered the low, dark corridor that led to the stairs, the old, crooked door slammed shut behind her.

Lynne and Neil had decided to get back into the pub game after a four-year break as civilians in Worcestershire. Lynne had tired quickly of work in the local Tesco. She missed pub life and the sense of community and satisfaction she got from running her own house. So, when they heard that a seventeenth-century inn ten minutes' walk from Land's End in Cornwall had come up, they decided that this would be the place where their new future would begin.

Lynne stepped into the pub and slipped behind the bar where her husband was busy cleaning the old till with a J-cloth.

'Dead chilly up there,' she said.

'Is it?' he said, stopping what he was doing and looking up at Lynne in her bright layers of woolly insulation. 'It's a lovely day though.' He thwacked the dirty cloth open and motioned towards the open door.

Lynne smiled at her husband, at his shaved head and wash-worn rugby shirt. She put her arm around his waist. He already looked absolutely at home, she thought. He looked perfect.

'How you getting on?' she asked.

'Fine, yeah,' he said. 'All good.'

'I still don't like it in that front bedroom,' she said. 'Can't we sleep in the other one?'

'Other one's too small,' he said.

'But it's freezing in there,' she said.

'It'll warm up,' Neil said, as he turned to look at the grey-haired, grey-eyed man who was sat at the bar. 'Can I get you another, mate?'

'Pint of Best, yeah,' he said, tipping the head of his empty pint towards them. 'Talking about that big room up there?' he said.

'Yeah, front one,' Neil said. 'Lynne's just unpacking in there. Says it's cold.'

'That's Annie's room,' the man said, nodding.

Neil and Lynne looked at him.

'Annie's room?' said Neil.

'Yeah. That's why you're cold in there. It's Annie's room. It's haunted. Hasn't nobody told you about Annie yet?'

'A ghost?' Lynne said, smiling. Poor old bugger, she thought. He's either trying to be friendly or he's trying it on. Either way, you've got to humour them.

'Yeah,' he said. 'We've all seen her. You not met Derek, yet? Old landlord?'

'No,' Neil said. 'Still a local, is he?'

'Oh, yeah,' the man said, nodding again and glancing at the ale tap. 'He'll be in soon. Derek'll tell you. That's Annie's room.'

Neil pulled the pump down quickly. A hard draft of beer shooshed into the fresh glass with an unexpected force, splashing up over Neil's hand.

'Well, Annie's just going to have to get used to us, isn't she?' Neil smiled. He finished pulling the pint, slowly now, and, when he was sure the old punter wasn't looking, gave Lynne a tickled wink.

That night, after lock-up, a huge wind reared up over Sennen. It was barracking the small window of the big, cold room upstairs as Neil climbed into bed next to Lynne. Safe and together under clean sheets, they chatted for a while over the noise, which excited yet quietly unnerved them. They talked, but they didn't mention the storm. They just chatted about the first night's takings and about how they missed their pets, who were still in kennels. Quickly, though, fatigue began to wash through them and Lynne leaned over to switch the light off. The moment the bulb died, the room became icy. Lynne pulled the covers up to her neck. She moved across to Neil for warmth. She put her arms around his middle and felt his weight rise and fall as his breathing lengthened and slowed. And then, he jolted.

'Neil?' Lynne said. 'You all right?'

He jolted again. His face twitched sharply as, outside, a new wave of wind arrived, fresh from the Atlantic, and crashed down onto the roof of the pub. Lynne looked over at the window as it rattled and realised the room was pitch-black, despite the street-light outside and the time her eyes had had to get used to the darkness. She could see nothing at all, as the temperature dived still further and the air took on a damp, musty smell. Neil made a small grunting sound and jolted again.

'Neil?' Lynne said again. 'You all right, love?'

'Nets,' he mumbled, from the deep of his nightmare.

'Neil?' Lynne moved closer and shivered and tried to make out the pattern of the wall or the shape of the big wardrobes against it. She couldn't. She was scared and freezing. She was confused. And she was shocked by the absolute darkness and the smell; it was horrible and it smelt so old and so wet and the black was so black and so thick and so much.

'Get the nets off,' Neil said. 'Get them off me. GET THE NETS OFF.'

Morning light came like an antidote. The room was real again – solid and uncomplicated, as another warm Cornish day began outside of the small sagging window that looked out towards Sennen Bay.

'Come on, love,' said Lynne, squeezing Neil's shoulder. 'Best get up.'

'Yep,' he said, sniffing and clearing his nose and throat.

'Did you have a good night's sleep?' Lynne said, catching his eye.

'Yeah, fine.' He smiled. He raised his arm and scratched at nothing on the back of his neck. With that expansive movement, he tried to distract her from the question, and from the answer, and from the fact that he wouldn't meet her look.

'I'll put the kettle on.'

All day, the bedroom was freezing. And every time Lynne left it and walked down the low, dark corridor that led off it, the door would close by itself behind her, even though it was stiff and deformed with age and difficult to push shut by hand. She was up and down between the pub and the flat for most of the day. With all the work there were few moments for reflection, and cashing-up time came quickly. And, as the night slid over the land and the sea again, Lynne sat at the bar, watching Neil at the till.

'Do all right today?' she said.

'Yeah, think so,' said Neil. 'That old landlord's not been in yet – that Derek. Wouldn't mind having a chat with –'

Suddenly, Neil was still, his eyes frozen and wide.

'What was that?' he said.

He looked around. There was a silence.

'What was what, love?' said Lynne.

And then, it was gone. Neil relaxed and shrugged his shoulders and let a short, dismissive laugh come out of his nose and his skin was clammy all over and tight and puckered with goose pimples.

'Er ... nothing,' he said, and looked away. He carried on with his counting. The till area was the only part of the room that was lit now, and Lynne was suddenly overwhelmingly aware of the depth and alien unfamiliarity of the shadows that surrounded her. Suddenly, their new home didn't feel like their home any more.

'Can't get the telly to work in the bedroom,' she said, rubbing her arm. 'It's weird. It'll work in every single room but the bedroom. Just can't get reception in there.'

'Maybe it's Annie,' said Neil, and they both laughed, watching each other as they did so.

Half an hour later, in the large room with the small window, Lynne leaned over Neil and turned the light off. She lay back, moved to pull the quilt up to her shoulders, and then it came into the room with them. It came closer and closer and then a huge, invisible weight fell on her and pushed her into the bed. She strained, managing to push her head up just an inch, but it was slammed back down again and then she was choking, being pushed, pushed down into the bed as the darkness stole the room again and the smell swept in, old and musty and dirty and damp. Lynne struggled as she pushed up, up and tried to call for Neil and she could feel the warmth of his sleeping body on her leg and the freezing, bitter air on her face as she lay there, straining, alert, terrified, heavy, helpless, scared. And then, she saw it. She saw it, first a movement, then a shape and then, there it was, a black shadow, a tall, black shadow moving out from the wall and coming towards her as she struggled and fought and tried to lift her arms in the oily, freezing, stinking blackness. Lynne tried to take breath. She lifted her chin higher and pushed it up and up, a bit higher and there was water, there was water, and it was coming up, higher and higher. She tried to call out and she felt like she was drowning, she was drowning, she was pinned down in the sea and the salty freezing water rose and rose and it came and it lapped and broke around her chin and her nose and she tried to scream out, *No, No, God, Help, No.* And then the shadow came and it moved in nearer and she could see it was a woman, a woman in a dark shawl that was tucked in through her arms. And it walked as Lynne lifted her chin and tried to breathe and then Annie was so close that Lynne could see that her clothes were made of black cloth and the cloth was rough like hessian and she could see the moonlight on the water and she was drained and frightened and struggling, fighting and praying and then the smell came in stronger. The foul, ancient stink blew in fierce and low and Lynne tried again and again to fight her way free as the water lapped and broke and rose and she was tired, so tired, the exhaustion dry and ripping through her muscles and the sea came in over her chin and her mouth and she could see the lights far away on the shore as freezing water rolled into her ears and over her head. *I'm not frightened*, she told herself desperately as the water fell into her mouth and pushed up her nose and she began to drown. *I'm not frightened, I'm not frightened.*

Then, the sea was gone, and she was back in the room again. She could move but she was scared and stiff. She watched the woman walk around the bed and stand over Neil, who started jolting and twitching and crying and begging. 'Get the nets off,' he was crying. 'Get them off me, get them off me.'

And suddenly, Neil woke up.

'Are you all right?' Lynne said to him, as her lungs thirstily pulled in breath.

'No,' he said quietly. 'Not really.'

They could both feel it, oppressive and strange and terrifying and lurking in the cold. They could both feel it, and they didn't want to talk about it.

'If you want to sleep with the light on, Neil,' she said, 'sleep with the light on.'

Neil rolled over, and switched it on. The walls were instantly washed clean as the bulb worked its white magic. And Lynne and Neil slept fitfully until the morning came again.

The next day, their pets arrived from kennels. It was their happiest day yet and Lynne felt reassured, that night, as they prepared to turn the light off again. But as soon as the darkness returned, Rosie, the old ginger cat, screamed and fought out and spat. And then Lynne couldn't move, the blackness and the ancient, musty smell came in, and Lynne was pinned down again, drowning in the sea, as Neil was flailing in his nightmare of the nets and the cat was on her hind legs hissing and pawing at the air and outside the dogs were barking and spinning and crying and then, Annie came again. Lynne thought, *I'm not frightened of you*, as Rosie hissed and raged and the water rose. *I'm not frightened of you.*

The next morning, Lynne and Neil got up and found that their cats had disappeared. They searched the flat, the pub and the cellar for hours. They called and whistled and rustled packets of food. Then, in a quiet moment, Neil heard a muffled, curling sob coming from the bedroom. They hadn't looked in there. There'd been no point – they'd shut the door behind them when they'd left that morning, and they knew the room was empty. And yet, he heard it. A cat's cry coming from the big, cold bedroom at the front. *How could they be in there?*

Neil walked through the old door, which moaned as it opened, and looked inside. He saw nothing but the white walls and the wardrobes and the small old window that looked out towards the bay. And then he heard it again.

The wardrobe.

'Lynne!' he shouted and went to the closet. He turned the old, steel key and he opened the door. Just as Lynne came in, she saw their black and white cat leap out of the wardrobe, scream and dart away.

'Shit!' said Neil. 'How the fuck –'

'Neil ... ' said Lynne, and there was more crying now. This time, from one of the two drawers underneath the wardrobe. Neil looked at Lynne. Lynne looked back. They breathed and Neil went, with a shaking hand, to open the small drawer. They were near to the edge of things now, all right. The edge of tears and of panic and of dread and of nothing they could ever hope to possibly understand. Neil pulled the upper drawer, and it opened with an old, wooden whine. And there was Tabs, curled inside it, petrified and trembling. Neil lifted out her warm, stiff body and passed her to Lynne. He opened the next drawer. And there was Rosie, lying on her side, shaking, weeping and trapped.

'Fucking hell,' he said, and looked at Lynne. 'Fucking hell, what's going on?'

'I don't know,' Lynne said.

But, by now, she did know. By now, she knew all about it.

'I want to move to the other bedroom,' he said.

'OK,' Lynne said, as the cats fled down the corridor, 'but before we do, I think there's something we need to talk about.'

'What?' Neil said.

'You're frightened of the ghost, aren't you?' she said.

'How did you know?' he said.

'Because the nets you're dreaming of, the ones that you think are being thrown onto you,' she began.

'Yeah?' said Neil.

'They're not. They're throwing them onto Annie. They're throwing them onto her to get her out.'

'What do you mean?' said Neil.

'Derek came in last night,' she said, 'and I told him what had been happening and he said, "That's Annie. She wouldn't do anything to harm you. She's just trying to tell you how she died. It's all about it on the door." And I said, "What door?" And he showed me. You know the door that leads up here, from the pub? The one that we've kept open since we moved in?'

'Yeah?' Neil said.

'It's got a plaque on it. On the other side of it, the side that's been against the wall all this time. Come and see.'

They walked through the cold corridor, down the stairs and into the pub, and Lynne kicked the doorstop away. The door closed and Neil read the hand-painted sign that had been hidden on the other side:

> *Former landlady Ann Treeve presided over smuggling and wrecking operations, together with the local parson, until turning Queen's evidence against Dionysius Williams a Sennen farmer (a smuggling agent) who then served a long prison sentence. For Annie's 'service' to the Crown she was staked out on Sennen beach and drowned by the incoming tide. Her body was laid out in the large upstairs room in this inn, prior to the burial in an unmarked grave for fear of retribution by way of grave robbers.*

I look at Lynne standing in the kitchen in front of me, as the autumn wind picks up outside and she stirs our cups of tea.

'Christ,' I say, 'that's weird.'

Just then, behind us, Neil walks along the corridor towards the lounge. 'Scary's what it bloody is,' he says as he passes.

Lynne smiles at me. 'Neil doesn't like to talk about it.' She hands me the warm mug. 'You shut the bedroom door when you're in there tonight,' she says, 'and you'll see. She is about tonight. You can feel her. It's almost like you're being watched. Especially walking along the passageway outside the room. She doesn't like being talked about, I don't think. Things seem to happen when she's talked about. And Rosie's behaving very strangely. She's batting the air and there's nothing there.'

I read about Neil and Lynne's experiences on the online version of their local paper, the *Cornishman*, during my lunch-break, last week. I got two paragraphs into the report, tracked down their phone number, called Lynne up and asked if I could stay the night in Annie's room. Once she'd said yes, I picked up my sandwich again and finished reading the article. I got to the bit about the drowning and the nets, then the bit about the pets who got shut in the wardrobe, and then the bit about the body being laid out in the room. This was after all the arrangements had been made. After Lynne had said, 'Are you sure about this?' and after I had laughed cheerfully and said, 'Oh, yes! Honestly, it'll be fine.'

'Lots of people have got stories about her,' Lynne says. 'There's Derek, who you've met, and a couple of weeks ago, we met another previous landlord and landlady. They run another pub now, not too far away from here. We went in and introduced ourselves and they said, "How you getting on with Annie?" It was one of the first things they said.'

Half an hour ago, I did have a chat with Derek, downstairs in the bar. A soft, melancholic and slightly wary man, he told me that he'd seen Annie when he was living here, and that when there was 'aggravation or upset' in the pub, pictures would fly off the walls and bottles would be seen moving slowly across the bar and smashing onto the floor.

'When I saw it for the first time, I couldn't believe it, like,' he told me. 'It almost always happened when there was problems. She don't like upset. She got pegged out on the beach, you know. They put her in an unmarked grave. That's the reason. She hasn't been left to rest properly.'

As well as the moving objects, the apparition and the horrific night terrors, Lynne says some people have had their hair pulled in the flat. When I ask if they've had problems with power-drain she nods. They've a cat-shaped clock, she tells me, that didn't need its batteries changed for years. Until they moved here. They've been replaced six or seven times now, in just a few months.

It's bedtime in the last village in England. I leave Lynne to join her husband and I walk back towards the room. I enter a small lobby area that's decorated with photographs and cluttered with stray bits of pub. An old menu board leans against the wall. Green bottle crates are piled by the banisters and next to them there's a tatty stack of wooden chairs. In the middle of all this, there are pets going crazy. One cat's cowering under a chair. The dog is barking. Another cat bombs across the lobby, towards the safety of its owners, its tail low to the carpet. I'm troubled by all this, as I walk through the furry mêlée. I'd be troubled by it even if I hadn't just been told that disturbing story. As I enter the corridor that leads to the bedroom, I grip my tape player and notepad tightly, deep in ponder. And then, I barely notice it, but something happens. The part of my brain that isn't trying to find an explanation for the spinning animals hears movement in the bedroom. But because I'm completely preoccupied, that small mind-part just says 'there's someone in the bedroom' and I carry on walking the next few steps, down the old corridor, go through the door and look up to see who it is. But there's no one there. Just the bed and the walls and my stuff on the floor and the shadows that the dim lightbulb is conjuring. I was so sure there would be a person in the room – the sound was so mundane, so ordinary and clear – that in those brief seconds, it didn't even occur to me that the room would be empty.

I stall in the doorway, startled. I'm scared. Only my neck is working, and that only in small, jerky movements. I get the sensation that I mustn't move, in case I draw the attention of something lurking. It's as if a sudden action from me might provoke some unknowable retaliation. I sniff unnecessarily, then clear my throat as a warm-up to breaking the thick silence. I hum a couple of notes, pause for a second and enter the room.

After I've lain, slowly, on the bed, I'm relieved to see a wet black nose peer around the door. The old dog has come to see me, followed by the ginger cat. I sit up and stroke them, as the relief works on me like warm oil and loosens up all the tense bits. There can't be anything nasty in here, I think, if the animals have chosen to come in.

As I play with the pets, I take the opportunity to look around me. The foot of my bed faces the small window and the old door is to my left. To my right is the stone wall from which Annie apparently steps out. Opposite that is the wardrobe, with two drawers under it, the one that the cats were found in. There's not a lot else here, except a row of Lynne's shoes, a picture of a fishing boat, a cheap plastic bedside light and a cat-shaped clock on the window sill, which has stopped.

I try to distribute attention between the animals fairly. It's not easy. Every time I do the dog, the cat starts miaowing. Then, when I stroke its bony back, the dog gives me the big dewy-eyed trick and I'm compelled to think again. I experiment, for a while, at doing them both simultaneously, but I can't co-ordinate my hands properly. Then, as I'm trying to work out a new system, the dog turns towards Annie's wall. It looks at something for a second, and runs out of the room. The cat sees it, too. And it's gone so fast that it skids around the corner.

I look after them. The door's open a few inches and the light from the corridor, which Neil and Lynne insist on keeping on all the time, glows dully through the gap. I prepare myself for the shock I'll get when it closes by itself. I wait and watch the door. It stays absolutely still.

Half an hour later, I've finally willed up the courage to shut it myself. The door is closed and everything is quiet and waiting. It feels as if the walls are watching me, as if this room has all the power and I've placed myself, willingly, between its teeth. Lying there, under my thin sheet, with the wind big and hostile and banging outside, I feel tiny and lonely and scared.

I take a breath, lean over and switch off the light. Then I lie back and wait.

Nothing.

I decide to speak. 'If you are here, Annie,' I say in a small voice, 'will you let me know?'

Silence.

'Please,' I whisper.

I look in the direction of the window, back at the wall, and at the dead lightbulb hanging from the ceiling. I look at the wardrobe. It seems to hang malevolently in the air, like a skinned cadaver on a butcher's hook.

Suddenly, a shrill scream shatters the dark air.

It's my phone. I scrabble out of the sheet for it, and retrieve it from the tangle of my trousers that are lying in a crump on the floor.

'Hello?'

'It's me,' says Farrah.

'What's the matter?'

'I was on the sofa and something weird happened.'

'What happened?'

'There was this laughing all around, like mischievous children, and then it felt like I was coming out of my body, like an out-of-body thing, like I was floating. I was off the couch completely and it was like the wall with the picture on it was where the floor should be and the room was all in weird proportions. I'm scared. What are you doing?'

'Nothing, Far,' I say. 'Are you all right?'

'I don't know. I'm scared. Why do I always get stuff when you're out bloody ghosthunting?'

'Um,' I say, 'it must have been a dream.'

'I know,' she says. 'It didn't feel like one.'

'You should just go to bed, Far,' I say. 'I'm sure it was just a dream.'

'But I didn't wake up, or anything,' she says. 'It was just like everything became normal again, like I went back in my body.'

'Are you OK now?'

'I'm scared ... I'll go to bed.'

'Goodnight,' I say. 'Call me if anything else happens. Love you.'

'Goodnight,' she says. 'Love you.'

I get back, slowly, under the sheets, where I lie still and wait for Annie.

16

'And when they die, they'll get a big surprise'

'You can't just expect things to happen to order,' Maurice tells me, the next day.

I lean back on the sofa for a moment, glance at the ceiling and then hunch forwards, towards him.

'But every time Lynne went in the room, the door slammed behind her,' I say, my arms gesticulating helplessly. 'Every single time.'

'Listen,' says Maurice, pulling himself out of the throat of the plump, flowery armchair that's trying to swallow him. 'As I believe I have told you before, paranormal activity does not act to command. It never happens when you think it's going to happen. I mean, to go to one of these situations and expect something to happen, well, it's quite rare. It's quite rare, I can assure you.'

'But every night,' I say, 'every night they slept in there, something happened. And Annie's meant to act up when there's change or strangers, so I thought I'd have a good chance.'

'Well, that's fine,' he says. 'That would be a fine situation. If anything had happened. But it didn't, did it?'

I'm not sure whether the world's greatest ghosthunter is flabbergasted by my naivety, or amused by it or irritated. That doesn't stop me, though. I'm over-tired and under-prepared and I've momentarily lost control of my talking.

'But why?' I say. 'Why does nothing ever happen when you want it to?'

'You want an answer?' he says.

'Yes.'

Maurice looks at me, his eyes watery, granite and pale. 'Bloody-minded.' He punctuates the words with a small nod for emphasis. 'Bloody. Minded. It's always like that. Always. It's as though it's saying, "Sod you." Bloody-minded, that's what I say.'

I wonder if Annie's non-appearance last night could have had anything to do with what they call the Experimenter Effect. This is the theory that the experimenter himself might affect what happens, just by being present. Nothing ever occurs, they say, in front of a sceptical experimenter. This may sound dubious, a suspiciously convenient excuse the ghost-convinced can give the protesting non-believer to shut him up, but something similar to this has been proved to happen in science – in quantum physics.

What I'm about to tell you is, quite possibly, at least four times more astounding than anything I've come across in my search so far. And it is real. Demonstrably, scientifically, unarguably real.

They have found – and nobody can explain this – that atoms will behave in one way if a human is watching them and in a completely different way if not. It's called the 'Two Slit Trick'. For one reason or another, some advanced science types with a spare morning to fill decided to fire some atoms through a sheet that had two slits in it. They were expecting the particles to act like sand and pile up in two equal-sized atom-stacks behind each slit, on the other side. But they didn't. Each atom actually behaved in a way that suggested it had gone through both slits at the same time. Which is, of course, impossible. Or paranormal. Or supernatural.

Wanting to reveal the process behind this unexpected atomic weirdness, they installed a tiny camera in their set-up and filmed the atoms in flight. But when they watched the tapes, the atoms actually did start to behave like grains of sand, and piled up neatly at the other side of the slits. Then, when the scientists switched the camera off and repeated the test, they started behaving mysteriously again. It was as if the atoms didn't want to be seen acting in a magical way, as if they wanted to keep their secrets.

And I thought that was just like ghosts.

So I decided to read more, and I found out about String Theory, the cutting-edge quantum-related idea that everything is made out of tiny, vibrating strings. If this theory is correct (and, so far, many of the world's leading brains think that it is) it means that there aren't four dimensions, as we have been led to believe. No. There are, in fact, eleven.

On one level, String Theory explains how quantum physics works in the same universe as Newtonian physics. Which is excellent. But it also creates a big new problem, because the way it pans out means there are whole slices of existence that we weren't previously aware of. Humans only have the equipment to detect four dimensions, so that's all we can ever possibly know. According to String Theory, the others are all around us, but because we only have a nose, eyes, ears, tongue and skin to receive information about the world, we have no way of experiencing them.

So, I was thinking: could ghosts or the afterlife exist in one of these other seven dimensions? Could quantum scientists have found theoretical evidence of the same invisible world that I've found physical and anecdotal evidence of? Have they found out where ghosts come from? Could these spirits be cross-dimensional leakage? And could the 'Two Slit Trick' explain their science-proof behaviour?

Maurice rubs his chin and looks away for a second to compose himself. 'Right, let me tell you again. There are no definite answers to why these things happen. If you're looking for answers, you aren't going to get any. You'll only get answers from charlatans. A lot of my colleagues believe there is a connection between quantum physics and paranormal activity. It is quite possible that paranormal activity, or the afterlife, might exist in the quantum sphere. But we don't know. Everybody's fishing and nobody's come up with anything yet. I mean, who knows? Quantum physics, the Experimenter Effect – they may have a great deal to do with what happens.'

He pauses again and frowns. 'I can tell you what does have an effect, though: the stress situation. There's absolutely no doubt about that at all.'

By now I've almost completely lost control over which thoughts are staying in my mind and which ones are coming out of my mouth. Part of the reason for this is that I'm exhausted after last night and the pre-dawn rise that I had to endure in order to make it to Muswell Hill on time for our meeting. But most of all, it's because I'm stalling. You see, I've actually come to see Maurice to confront him about my Enfield research. But I'm worried that when I remove the Anita Gregory Dossier of Doubt from my bag, the ghosthunter might get up and throw me right through his pretty bay windows. I can't get the words of the SPR man out of my head: *'If Maurice Grosse hears you mention her name, he'll take your head off.'*

'I've investigated hundreds of poltergeist cases,' Maurice is telling me. 'Literally hundreds. And I've never been to one where there hasn't been a very strong element of stress. They've either got money problems or sexual problems or family problems or some bad problem that is causing a lot of trouble. It's quite remarkable.'

I'd forgotten how much Maurice has shrunk. I was properly taken aback when he opened the door. Since we last met, I've looked many times at his photograph in my copy of *This House is Haunted*. In times of weary fatigue, I have studied it for strength. It is an excellent portrait of a man who means business. A strapping, besuited professional, caught seemingly unawares, removing the keys from the ignition of his shiny red Jaguar. But he looks tiny now. Pale and slight and careful. Until, that is, he starts to speak. If age has taken something of Maurice's physical presence it hasn't stolen any of his spirit. He's still fierce and sharp and quick as a switchblade.

'This one you went to last night,' he says, 'was there any stress?'

'I don't know,' I say.

He glowers at me over the top of his moustache. 'Then you didn't ask the right questions, did you?'

This 'stress effect' brings me back to the idea that ghosts need to gather some sort of energy in a haunted situation. From the flickering lights to the draining batteries to the Ouija boards that, supposedly, work by letting a spirit use human energy, it does all seem to point to the suggestion that ghosts need to feed in order to be. And perhaps they can feed on emotional energy, like fear or stress, as well as other types.

'You're talking about energy,' says Maurice, 'but you don't even know what that energy is. Nobody does. Guy Playfair has told me about a poltergeist case in Brazil where a car was literally lifted up and thrown a hundred yards. If that's true, what sort of energy was being used there? And, in any case, we're not just talking about an unknown energy that's moving things; it's affecting people as well. Like in cases of possession. If it can do those things, you might say that it's capable of anything.'

Maurice has his theories about telepathy and psychokinesis. He thinks these are real human attributes that could explain away much apparent poltergeist behaviour. And he's right, of course. Just as hypnogogic states might explain some night-terrors, the Stone Tape theory some apparitions, advanced body-language skills some mediums, coincidence some crisis apparitions, rogue radio waves some EVP, and hoaxers, mental illness and wishful thinking the rest. Nothing, however, explains it all. Except, that is, the afterlife theory.

'Well, yes,' he says, nodding sagely, 'there's an enormous amount of evidence for the afterlife. But how much poltergeist activity has to do with it, I'm not quite sure. The only thing in the Enfield case was that he identified himself in the voice and said who he was.'

'This is Bill,' I say, 'the guy who died in the house?'

'That's right,' Maurice says. 'Everything pointed to the fact that it was that man.'

Whilst the Enfield investigation was ongoing, Maurice was interviewed on the radio. During the broadcast, he played a tape of the voice. Soon afterwards, a man who'd heard the show contacted Grosse and told him that he'd lived at the house and that the sound on his tape was the voice of his father, Bill.

'And he did tell me some facts that we didn't, at that point, know,' says Maurice. 'He also told us how he died, and he was exactly right.'

So was it the spirit of Bill that was haunting the house? Was Bill the entity *and* the voice? Or did Janet speak in that gruff growl because she was on some sort of extreme attention trip? Anita Gregory thought so, I mention, very quietly.

There's a small, dense silence. I sit and wait in slow motion.

'Anita Gregory,' Maurice says. He lets out a small sigh and glances out of his window. A car passes by. 'What a problem she was.'

'I was trying to get hold of her thesis,' I say, 'but it was placed out of bounds.'

'That's because we threatened her with legal action,' says Maurice.

'I understand you took legal action against a journalist, too?'

'Oh yes,' he says. 'I've got the apology upstairs.'

'So, did you prove the Enfield case was true in a court of law?'

'No,' he says, 'all I did was get my solicitors to write to them and got an about-turn. I'd just like to tell you – I've never been called a liar. Ever.'

'OK,' I say, 'but why did you force the SPR to place Anita's thesis out of bounds?'

'I'll tell you what happened,' says Maurice. 'She came to the house maybe three, four times. One day, she was writing up her notes in a book, similar to the one you've got there. I was also writing up my notes in a book. Now, she went home one evening, and when I opened my book, I realised that she'd got my book and I'd got hers. Now, when I read those notes, I thought, this is a very disturbed person. She's writing, "I don't know what the hell's going on here" and "It doesn't make any sense" and so on and so forth. And yet, when she did that thesis, she said, "Oh, this definitely didn't happen and this definitely didn't happen." If I'd have been clever, I'd have kept that book. But I didn't. I gave it back the next morning.'

'Still, though,' I say, 'to threaten her with legal action – that's quite an aggressive act.'

'Well, no. I wasn't too worried about it,' he says. 'But Guy got very upset. Well, Guy's an author, you see, and it was questioning his book and his honesty. I was naive enough to think that it didn't matter a tuppenny damn.'

'So, I take it that she wrote that the Enfield case was faked?'

'Oh yes,' he says. 'Her thesis was very bad indeed. John Beloff told me afterwards, he said, "That influenced me at the time, but now I believe everything you said."'

I shift in my seat. 'That's funny,' I say. 'John Beloff wrote to me and said it was the kids acting up.'

There's a small silence. I wait. Maurice smiles. He rubs his chin.

'Is that what he said, now?' he says. 'Well, he must be getting old. Because it's not very long ago he said to me, he said, to my face, he said, "What I've said before about the Enfield case – I retract it all."'

'But,' I begin again, 'I wrote to him asking his opinion on the Enfield case and he said that he was of the opinion that it was the kids.'

'I bet if I confronted him face to face he wouldn't say that.'

'But –' I say.

Then Maurice's tone changes. 'Listen, I've had to put up with all this nonsense ... ' He pauses. 'Look, I don't care if ... let me put it like this. I don't care a tuppenny damn about the sceptics. I've spent my whole time, my career in parapsychology, trying to produce evidence. Now, can you point to anyone who has produced better evidence than I have?'

'No,' I say.

'Then, that's the answer to your story. All right?'

But it's not all right. Not yet. I read Anita's claim that the neighbour, Peggy Nottingham, said that what was going on at the end was a 'load of nonsense'.

'Well, she told me she said no such thing,' he says.

'But what about the WPC who witnessed the moving chair? Anita said she thought that was a trick.'

'Well, ask her how the trick was done,' he says. 'I've got her statement upstairs.'

In fact, there's no need for Maurice to fetch me the statement. I've already seen the transcript of an interview that WPC Heeps gave a director from BBC Scotland at the time of the haunting. I found it in the SPR journals. Her exact words were: 'The chair was by the sofa, and I looked at the chair and I noticed it shook slightly. I can't explain it any better. It came off the floor, oh, nearly a half-inch, I should say, and I saw it slide off to the right about three and a half to four feet before it came to rest. I'm absolutely convinced that no one in that room touched that chair or went anywhere near it when it moved. Absolutely convinced.'

'So, was Anita lying about that?' I ask.

'You want my honest opinion?'

'Yes.'

'I don't want to speak ill of the dead.'

'I know she's dead,' I say, 'but ... '

There's a long, driving silence. I look at Maurice. Maurice looks at me.

'She was a liar,' he says. 'No question about it.'

With that, I decide to put the dossier away. Even if you do accept some of Anita's concerns, I still think it's unreasonable to believe the whole eighteen-month episode was a hoax. I'm convinced that an eleven-year-old couldn't fool a mother of four, her neighbours and all those journalists and miscellaneous strangers that a ghost was causing chaos, when it was actually her. Nobody could. If you, as a child, threw a Lego brick at a house-guest's head and blamed it on a poltergeist, there wouldn't be a mother on earth that would fall for it. Not even once. And besides, why would she want to make it up? The haunting was an exhausting and miserable trial for the whole family and it led to Mrs Hodgson having a nervous breakdown, to her brother suffering years of bullying and to Janet herself moving away from the area. She'd have to have been crazy to willingly inflict that on everyone – and we know she wasn't crazy, the team at the Maudsley confirmed that much.

And then there are the parallels with other cases I've come across. Deborah Carven also complained of terrifying reflections in the window. Tim Laverty experienced hot stones falling from the air during his school trip to Michelham Priory. And, just like Kathy, Debbie and Loping Coyote, Janet would 'go under' and have no memory of what happened when she came back up again. And then there are the indismissable similarities with other poltergeist cases from around the world and centuries. Almost every item in Canon Michael Perry's checklist has been checked. So really, if we're going to conclude that Janet – possibly with the help of Rose – was responsible for an almighty hoax, as well as being extremely well educated (they weren't), they'd have to have been expert magicians, ventriloquists and liars. And unless during my evening with Janet I witnessed yet another astonishing performance, I just cannot believe that sad, faltering person was the evil architect behind the Enfield haunting.

In the end, reading through Anita's letters, I get the sense that she actually had more of an issue with Maurice and Guy's scientific techniques than she did with the authenticity of much of the phenomena. It seems to me that Maurice and Guy weren't logging and recording things in the way that she thought a professional should, and they might have felt irritated when this was pointed out in the SPR journal, which was why they were over-ravenous in their anti-Anita actions. In fact, in an article called 'Investigating Macro-Physical Phenomena', that Anita wrote for the *Parapsychology Review* in 1982, she repeats her doubts, but this time concedes that 'there is nevertheless *some* good evidence and testimony'.

But the one aspect of the Enfield haunting that Anita did absolutely refuse to accept, though, was the voice. Anita had a monstrous issue with the voice. So, was Janet putting it on? I think it's possible. But, to be honest, I don't think it matters.

I think the voice is a red herring. The tapping, the furniture moving, the doors slamming, the hot stones – all these things were witnessed by many people. Even Janet's levitating was witnessed by two locals. These things, taken in the context of their consistency with historical hauntings, well, that's good enough for me. You'd have thought that would be good enough for anyone. But it probably won't even be approaching the outskirts of good enough for the sceptics.

The problem, I found, with turning fully sceptical is that to really pull it off you've got to stitch your eyes shut, pump rubber glue down your ears and say, 'I don't care what anybody else says. They can throw whatever they like at me but it won't make any difference because I already know the truth.'

The other morning, as I was getting ready for work, I heard a scientist on Radio 4 telling the country with a satisfied chuckle that there is no afterlife. I paused in mid-air with my sock pulled half on. How does he know? Logically, that statement could not be based on science. It's pure belief. Even if someone *had* managed to properly discount all the paranormal evidence there is, well, that still wouldn't prove death is the absolute end. All this made me realise that to be a hard sceptic you have to start with the belief and *then* look at the evidence. Which, often, you have to twist and squash and smear to make it fit your point. And it's this partisan, almost political approach to the subject which leads to the ridiculousness of sceptics trying to prove their point by replicating paranormal phenomena with magic tricks. To me, that makes no sense. You don't try to explain snow by faking a blizzard. Just because you can forge a £50 note, it doesn't mean that all the red bills are snide. Another one of their tricks is to say, 'Well, it's just not very *likely*, is it?' Well, really, that depends on who you ask. And how confident you are that twenty-first-century humans already know everything there is to know about all of existence.

To take the sceptical stance, you need to just *know* somewhere deep in your soul that you're right. In other words, to be a true, fundamentalist hardcore disbeliever, you have to have faith. And I find faith difficult. Because it's that closed attitude to life's mysteries that made me want to walk out of my R.E. A-level lessons. The hard sceptics, you see, are just the same as the priests. Because they offer absolutes, they give answers – and Maurice is absolutely right about that: you only get answers from charlatans.

That morning, as I stood in my bedroom with half a sock on, I wondered why the sceptical faithful just cannot accept that there might be more after life than the void. It's as if they've got a blockage.

'A blockage!' says Maurice. 'Yes. They've got a blockage.'

'Is it because they're scared?' I ask.

'I think that some people don't like the idea of an afterlife because of the life they're leading now,' says Maurice. 'This question of heaven and hell is much more deeply ingrained in people's psyche than they think. Especially if they've got a Catholic background. Another one is if they've been brought up with no spiritual background whatsoever. You have to understand that when you're talking about sceptics, you're talking about a very special breed of people. It's … it's … '

'It's Luddite,' I say.

'That's it!' Maurice roars. 'That's the word! Absolutely! Here, do you want to hear a tape of Janet speaking in the voice?'

'Have you got one?'

'Does that thing play back?' he says, looking at my Dictaphone.

Maurice goes upstairs, and returns with an old cassette tape. I slip it into the player and press play. It hisses. And then a noticeably younger-sounding Maurice says: 'You understand that you shouldn't really be in this house?'

A faint knocking responds. Then, the voices of the Hodgson family can be heard. More knocking.

'Shh!' he says. 'It's now doing the rat-tat-a-tat-tat. Now, I'm going to ask you a question. Are you having a game with me?'

There's a noise of something. And then …

'Aaarghh!' says a young girl. 'Oh!' says an older woman. 'Oh! Crikey!' says Maurice, all at the same time.

Then Maurice speaks. He's out of breath, excited.

'As I asked the question, "Are you having a game with me?" it threw the cardboard box and the pillow right at my face. Thank you very much, that was a very good answer.'

We forward the tape.

'Say, "Doctor Beloff",' Maurice says. 'Come on.'

'Doctor! Beloff!' it shouts.

The voice. My first instinct is to try not to smile. It just sounds stupid. This is a tape of an eleven-year-old putting on an old man's voice. But then, I think, let's just say that a ghost *was* using the physical equipment it found inside an eleven-year-old's gizzard to speak. What would it sound like? Well, probably, just like an eleven-year-old girl putting on an old man's voice.

But I still don't think that belief in these voices is essential so my 'wows' are mostly manners-based. Until Janet's tone suddenly barks. I bend down towards the tape slightly.

' ... so you can't shoot me,' it says.

'How can we shoot you if we can't see you, Bill?' says Guy Playfair.

'BY PRAYING! To God.'

The voice growls along, like an oil drum being dragged over gravel, and then, without warning, it leaps into a quick, sudden yelp and then goes back down again. Without hesitation I know where I've heard this before. It's precisely the same pattern of sound as the EVP Lou recorded in Kathy's house and at the Carvens'. Really, precisely, completely, exactly. The timbre of the growl is identical, as is the timing and suddenness of the yelps. And it's the same in such an unusual, unexpected way. I stand there, open-faced, as a chill wave breaks over me. I tell Maurice.

'How strange,' he says. 'That's very interesting, what you told me. Very interesting. I've never heard that before.'

Then I talk about the cavernous glare, like a traction beam into purgatory, that came out of Janet's eyes, and how it was exactly the same cavernous glare that came out of Kathy's.

'And I wasn't looking out for it,' I say. 'I wasn't expecting it. And I've never felt that from anyone else.'

'Well, there you are, you see!' he says, and laughs at the expression of gentle horror that this connection has left on my face.

I'm stunned because the fact that this voice – which was recorded in a London suburb in 1978 – sounds the same as one recorded over twenty years later, on a different continent, makes complete sense. It adds up. Because if both these recordings actually *are* the sound of disembodied souls, they *would* sound the same – whether they were free-floating in the invisible world or whether they'd managed to disappear up the nostril and down the throat of a human with usable voice-amplifying equipment. They would sound the same. And they *do* sound the same. It all makes heavy, terrifying sense. And it *is* terrifying because it means ...

'When people say they believe in ghosts,' I say, when we've sat back down on the comfy chairs, 'I'm not really sure that they know what they're saying. I mean, the ramifications ... '

Maurice leans forwards and offers me a plate of biscuits as a soothing gesture.

'Of course,' he says, nodding.

Maurice understands.

'It's the most important subject there is.'

'Do you believe in heaven and hell?' I ask.

'I don't believe in hell,' he says, 'and I don't believe in heaven. I just believe there's something else, and that when they die, people will get a big surprise.'

'But all this research that you've done,' I say. 'Hasn't it changed the way you view reality?'

'Absolutely,' he says.

'How?'

'I can answer that simply,' he says.

For the first time, I detect a marvellous hint of wonder and vulnerability soften the steel in his eyes.

'I'm not sure what reality is any more.'

17

'Some really weird things'

To my right, a vacant seat. To my front, a small table that holds my mess of notes, pens, research papers, tapes, a tape recorder, a laptop and no cup of tea. To my left, the black miles of night-drenched Britain, streaming past outside the window. I'm rarely happier than at the outset of a long, quiet train journey, when I've found a forward-facing place to sit where nobody else can see me.

It's been several weeks now since my meeting with Maurice. Much has happened. I've made an inglorious return to Michelham Priory with the Ghost Club. All was entirely normal – nothing, as far as I was concerned, remotely para occurred. Paolo was there, doing his thing. And Philip Hutchinson was there, too. We shared a disenchanted grumble as the dawn leaked out over the wet, mist-freezing lawns. I've also been back to Newcastle to see Debbie and Trevor. I hired out Newcastle Keep myself so I could test out a couple of theories. Firstly, I decided that the yabbering assemblies that normally attend vigils are probably not that conducive to the coaxing or sensing of fleeting ghostly happenings. So I thought I'd arrange for as small a group as possible to sit in the still and quiet. And this time, Farrah came with me. As she'd kept experiencing freakish events while I was out searching, I thought she might have been more psychically inclined to encounter something than me. She wasn't. At just gone 3 a.m., she crept into an anteroom with a small gas fire and curled up on a wooden bench to fall asleep. None of the rest of us experienced anything remarkable either, unless you counted the bit when Debbie suddenly announced that she'd been shot in the face with a musket. *I can taste the blood in my mouth,* she said.

And something else has happened, too. Christmas has been and gone. Mother Nature has given the world a totally bastard present – a tsunami, in the Indian Ocean. A mere muscular twitch in the earth's physiology that has killed thousands. I pull myself up in my seat and let my fingers scramble through my notes for the newspaper cutting. Here it is: 'Domestic and foreign media have reported dozens of ghost sightings in tsunami-affected areas. The encounters have ranged from back-packers heard laughing and shouting on deserted beaches to holidaymakers hailing a cab to the airport, only to disappear along the way.'

I put the cutting down and look to my left. Rain has started to hit the window. Fine little drops are shivering, merging and streaming in thin streaks. I find a tape, in amongst my rubble, upon which I've written 'Michele Watkins', and slot it into my Dictaphone. We run into a tunnel and the hailstorm sound of the speeding inter-city is pushed back into the carriage. The train rocks and roars with an intimidating fury. I put my earphones in, push the play button and the sound of the bright, 44-year-old civil servant's voice begins:

'I'd always had some really weird things happen to me. The first big one was, oh God, this was twenty-odd years ago. I'd started a relationship and I just thought he was a really nice bloke, and I had a vivid dream about him one night. Even though I'd done nursing, I'd never seen somebody have an epileptic fit. But in the dream, he had his head in my lap and he had an epileptic fit. And I mean really bad. And it was a beautiful sunny day, and I was freaking out, because I had never seen it. And then, the same night, I dreamed that he already had a girl-friend and that she was dressed in red and she was a hairdresser. Quite strange. So the next time I saw him, we went for a walk in the park, sat down and chatted. And I said to him, "I had a funny dream about you the other night. One thing disturbed me – I might tell you later – but the other thing that disturbed me was that you have another girlfriend." Well, he just went white.'

I look over at the uniformed man, turn my tape off and say, 'Yes – can I have a tea, please?'

'Milk and sugar?'

'Yes, please.'

'One pound twenty.'

Christ.

'Thanks.'

With my drink artlessly delivered, the man pushes his silver trolley away. I lift the lid of my tea and stir the bag. The smell of softly melting plastic causes my mouth to fill slightly with a sprinkle of pre-vomitous spit. I swallow it back and look down at the collapsed landfill of information on the table in front of me. These pieces of paper contain yet more tiny puzzles to add to my larger one. They're pieces and pieces of pieces of a vastly complex supernatural mosaic. One of the bits recounts a case-report that was published in the *British Journal of Psychiatry* in 1994. Its authors, Anthony Hale and Narsimha Pinninti, describe their successful treatment of a twenty-two-year-old Hindu man who claimed that he was possessed by the spirit of an old woman who forced him to commit a series of escalating petty crimes. He ended up being sent to prison for hijacking a taxi and kidnapping its driver. The man told doctors that when he was about to be taken over by the spirit, a white fog would materialise and drift towards him. Hale and Pinninti's report described their patient as an 'intelligent, well-educated and insightful young man, westernised in his appearance and apparent outlook'. They said he gained nothing from his actions, not excitement, nor financial gain. And they could find no evidence of panic attacks, no history of mental illness or drug use. Then, they say they were 'disturbed' by a phone call from the prison chaplain who saw 'the ghost possess the patient in prison, seeing a descending cloud and an impression of a face alarmingly like a description of the dead woman given to us by the patient, of which the chaplain denied prior knowledge. Similar reports came from frightened cellmates. He and our hospital chaplain concurred on genuine possession.'

I look out of my window again, at the orange streetlights of some town suppurating out of the rushing, rain-soaked darkness. I pour a dose of sugar and then plop a drop or two of synthesised milk into my tea before rewinding my tape a little and pushing the play button again.

'" ... the other thing that disturbed me was that you have another girlfriend." Well, he just went white. And he said, "How do you know that?" I said, "I just had this dream. She's got blonde hair. She's dressed in red. And she had scissors in her hand. She was cutting somebody's hair. Is this true?" He said, "Yeah. It's true. I do have a girlfriend. But how the hell do you know she wears red and that she's blonde?" I said, "I don't know. It was in the dream." Then he said that she worked in such-and-such-a-hairdresser's in town and they all wore red overalls in there. Well, I was freaking out. And he had his head on my lap, at this point, and guess what happened? He had an epileptic fit. Right in front of me. A grand mal fit. Luckily, with my nursing experience, I knew what to do. The whole situation wasn't just a little bit like I dreamed. It was exactly the same. It just made me think, what was the dream? A warning? But that was ages ago, when I was about twenty-two. The first really big thing that happened to me that was to do with the afterlife was much more recently ...'

I switch the tape player off again and get up to find the toilet, leaning on the corner of a seat to steady myself as I stand. By the time I've returned from my trip to the piss- and tissue-speckled cubicle, the rain has quietened. We've sped through the worst of the squall and are now at full cruising speed. I sit down again, carefully, trying not to tip over my almost-empty beaker of tea.

There are one or two pieces of mosaic, in amongst the rubble in front of me, that sparkle brighter than many of the others, and yet, I'm not sure whether they will actually end up being part of the bigger ghost picture. One of them concerns a man called Dr Michael Persinger and his work in an area called 'Neurotheology'. Persinger's research centres on temporal lobe epilepsy (TLE), a rare condition in which sufferers experience intense religious visions. Some believe they're in the company of Jesus, Mary or Joseph. Some believe they actually *are* Jesus, Mary or Joseph. One poor bastard had a fit and was convinced he'd seen a 'green-skinned devil'. TLE victims' attacks are triggered when naturally-occurring electro-magnetic energy leans into their lobes with its lightning-fork fingers and short-circuits their brains. Scientists think that St Paul and Moses were both possibly TLE sufferers because accounts of their respective 'road to Damascus' and 'burning bush' incidents echo the symptoms of the condition almost precisely.

Dr Persinger, a behavioural psychologist who works out of Laurentian University in Ontario, tried to induce a TLE attack by making people wear a specially built 'Koren Helmet' that passed an electro-magnetic field across their temporal lobes. In the end, he didn't manage to create any new Bible heroes. But he did find that eighty per cent of his guinea pigs felt what he called a 'sensed presence' – the creeping feeling that they were being watched. All this made me wonder, for a time, if Dr Persinger was on the verge of solving ghosts. So, I asked him if I could pop over to Canada for a visit. Initially, he agreed. But, for some reason, when I asked if I could have a go on his ghost helmet, he stopped returning my emails.

Soon after this, I found out that scientists in Sweden have been trying to replicate Persinger's experiments – except, this time, the helmet-wearers were not warned that they were going to be exposed to magnetic fields. When the Swedes followed this all-important 'double blind protocol', they concluded that magnetism actually has no discernable effect at all. The argument still rumbles on between the continents.

All of the paranormal groups that I've been with so far have used EMF detectors to check for unusual strengths in the local magnetic fields because they think that peaks signal a spectre's presence. But, if you follow Dr Persinger's logic, it appears that they've got it the wrong way around. It's these EMF peaks that are actually *causing* some sensitive temporal lobes to jump themselves into having an apparent ghostly experience. And then, of course, there's Lou Gentile and David Vee. Could they be third stage temporal lobe epileptics? Could their demon sightings actually have been undiagnosed TLE fits?

The train slows a little and, as it does, a storm of mechanical noise rises up, thick and clanky, from the machinery beneath the carriage. We must be approaching a station. I put my earphones back in, close my eyes, rewind my tape a little and press play.

' ... ... first really big thing that happened to me that was to do with the afterlife was much more recently. I suppose it must have been about two years ago. One morning I felt somebody sitting on the end of my bed, but I couldn't see them properly. I just felt them. I thought my son had come in. I said, "Oh, Andrew, what is it? Are you all right?" and there was just silence. I sat up and I couldn't see anybody – I just felt this pressure on the bed. Anyway, I went back to sleep again and didn't think any more of it. And then another night, something similar happened. And then, one day I woke up and I just heard this person laughing at the end of the bed and I freaked. I mean, you can imagine. I just thought that somebody had broken into the house. I just went, "Who the hell are you?" And I heard this Glasgow voice going, "Oh, for goodness' sake, Michele, don't you know who I am?" I just thought, am I dreaming? What's going on? I mean, I was freaking out, absolutely freaking out. And I looked and it was a friend of mine's father. But he'd died, a year before. Ha! I can just see him now wearing this Pringle jumper! He said, "I've just come to tell you a few things." And I said, "Oh my God, it's five in the morning." I thought, I must be dreaming, but I wasn't. I was quite wide awake. And he was as real to me as you are now. At first, he did seem like an outline, and then he was much more solid. And he was sitting on the end of my bed!'

I switch the tape off again and think for a while, trying to piece together some more fragments. I pick up a print-out of an email from the sister of a friend of mine. Her young daughter had an 'invisible friend'. And this lady didn't think too much about it, bar all the wonderful things it promised about her child's blossoming creative imagination. And then, one afternoon, she overheard her daughter chatting away to her friend while she played in her bedroom. And she walked in and saw the now suddenly not-so-invisible friend, just for an instant, before it faded away into nothing. Apparently, she still hasn't recovered. And she remains so aggressively baffled about the experience that I haven't been able to persuade her to talk to me about it.

I switch my Dictaphone back on.

' ... Well, I was kind of thinking, what do I do? He wasn't someone threatening to me. He was somebody I knew really, really well. And I knew he was dead. So I thought, OK, bring it on, let's just see what happens. And my heart was like this. I said, "What are you doing?' and he said, "Oh, I've just realised that you're open to all these things that I never used to believe in" – because he was very against Spiritualism or anything ... he was really a down-to-earth Glasgow guy. He said, "I've got a few things to tell the family that you need to tell them." I said, "All right, give it to me." And he said, "Right. The wife. She's not well. She's got a cold. I don't want her to be going out. I've had enough of this. She keeps going out in the cold and she's going to get something like pneumonia." He said, "The family's got to do all the shopping and the cooking, and whatever." OK. "Right, tell Alan, thank you very much for the lovely flowers that you put on my grave. The roses were absolutely beautiful." I said, "Well, OK." And he said, "Tell Jimmy" – and I didn't know who the hell Jimmy was – "to watch himself when he's playing football because he gets too aggressive and he's going to hurt himself so badly that he won't have a career in football." Now I think to myself, why didn't I ask something profound, like what is heaven like? Is it a wonderful place to be? But you don't. I was just so gobsmacked I didn't know what to do. So, anyway, that was that. And then I went to visit the family ... '

I notice that we've stopped again, in one of the towns that all look alike and punctuate the long journey northwards. There's mangy pot plants, chipped paint on wrought ironwork, an unmanned Puccini coffee shed and, oh Christ – people. And they're getting on. Quickly, I switch my tape player off, shove a load of paper onto the seat next to me and pretend to be asleep. Only when I feel the train slowly pulling into motion do I open my eyes and pick the sheets back up. In amongst them are my notes on Near Death Experiences.

I was keen to find out about NDEs because, I reasoned, if it *had* been established that someone, at some point, had actually been up to something while their brains were officially off, then this would prove that the body and the mind are, as I discussed with Dr James Garvey during the summer, different things. And this, in turn, might mean that ghosts are possible. Then I heard an incredible story about an American called Pam Reynolds.

Pam woke up one morning to discover that she had a fragile sac of deadly liquid swelling up inside her head. She was rushed to the Barrow Neurological Institute in Phoenix for a radical operation to have the aneurysm removed from where it nestled, right underneath her brain stem. The only way the highly specialist medics could do this was to drain all the blood from her head, drill off the top of her skull and tease the gooey bag out of its nest. It was while her brain and body were in the medical equivalent of the deep-deep-freeze that Reynolds felt herself popping free of her mortal machine and floating up through the air. She remembers staring down at her body, looking at the tools the doctors were using, hearing a conversation between the surgeon and the nurses while they encountered a major problem with the procedure and were forced to have an emergency rethink.

When Pam came round, all the details of everything that she witnessed from her unfeasible perch up by the ceiling were confirmed as being absolutely accurate by her cardiologist, Dr

Michael Sabom, and her surgeon, Dr Robert Spetzler. Throughout Pam's operation, every single clinical sign was being monitored by machines. So it's official. It was lab conditions. Pam Reynolds' system was completely shut down. And what happened to her that afternoon, up by the oily light-fittings, should have been utterly impossible. Unless, that is, the mind and body are different things. Because if the mind *can* float free of its physical vehicle, then Pam's experience can be explained.

Physician Stuart Hameroff and his partner Dr Roger Penrose are world experts in the study of consciousness. And the work that they're doing now might end up changing the way we view existence for ever. Because they *do* think that the mind and the body are separate things. Their research has led them to believe that our souls exist on the tiniest, most fundamental level of the universe – the quantum level. The one that doesn't like being watched by humans.

There are things, I learned, called 'microtubules'. These minute contraptions live in the base of our brains and act as on-board computers, containing the information and processes that are the very essence of ourselves – our soul, in other words. But that's not the really incredible thing. The truly tectonic-rocking breakthrough that Hameroff and Penrose have made is this: when our systems shut down – when we pass away – the information that's held in our microtubules doesn't die. It can't, you see, because it's part of the quantum level, which is the most basic level in existence. It's the level on which the very fabric of the universe – matter, energy, space and time – exists. And, what's more, when they drift free of our microtubules, these little specks of soul don't separate and float apart: a process called quantum entanglement keeps them bunched together. So, if it's correct, this elegant nugget of extreme science does appear to show that the mind and the body *are* separate things – and that they *can* exist independently. Our brains, these men claim, do not *create* consciousness. They just channel it, like a television picking up a station.

All this might explain why people who have Near Death Experiences describe suddenly feeling 'at one with the universe' and having the radiant revelation that 'everything in existence is interconnected'. Because a soul that's been released from its body *does* suddenly become absorbed into the universe. And in quantum science, everything *is* interconnected – that's what makes it work. Just compare these traditional hippy sentiments with the views of a scientist like Victor Stenger, professor of physics and astronomy at the University of Hawaii, who says, 'The universe is one and we are one with it.' Strangely, and for the first time in human history, it would seem that the scientists and the druids are in total agreement.

And there's more to come from the frontier sciences. One of the problems with ghosts that Dr James the sceptical philosopher coughed up concerned the physics of a spirit moving things about. He said: *'If a ghost and a body are different stuff, then the question immediately arises, how does one thing affect the other? The kinds of causal interactions that we understand are things like billiard balls smacking into one another. So how can a thing that doesn't exist in space, like a ghost, have an effect on something that does?'*

I was thinking about this when I read about some experiments in consciousness that were set up by René Péoc'h and the Swiss Foundation Odier de Psycho-Physique. The experiments involved a robot, called a Tychoscope, and some baby chickens. The scientists 'imprinted' the robot's image onto the chicks by showing them a photo of it as they emerged from the shells. Once this was done, the birds thought the Tychoscope was their mother. When enough of the chicks had been imprinted like this, they were put in a cage, which was placed in an empty room, with the robot. When this happened, the scientists found that the robot's behaviour changed dramatically. Before the arrival of the birds, as you'd expect, the robot ranged about randomly, spending equal time in both halves of the room. But when the chicks were added – all of whom were crying out for the love and attention of their robot mum – it spent much more time in their half.

The astonished scientists noted that the chicks appeared to be mentally willing their mum to be near them. And it worked. Their motorised mother *was* drawn near.

These experiments, which are similar to other incredible ones that have been carried out at America's Princeton University, seem to provide absolute, verifiable examples of something apparently non-physical having a physical effect. And that thing is consciousness.

Now, just do some wild imagining with me for a moment ... Say a man called Bill has just died. His entangled quantum soul floats free from his body and – what if he suddenly finds that he can interact with the universe that he is now part of? He could pull an iron fireplace out of the wall, say, or throw a Brazilian car down the street or pull Janet out of bed every night. And that's not all. What's stopping Bill's freelance soul just floating on into another person's microtubules and temporarily taking control of them, because he's pissed off that he's dead? He could, if he wanted to, completely possess a still-living human. He could make her talk. And, while he's at it, he could have her demonstrate knowledge of subjects that she should know nothing about, like the manner of his death and the site of his burial. He could even, if the fancy took him, compel her to tell an interfering priest to 'shit off'.

I look to my left. That last town must have been small, because all evidence of human habitation has disappeared from my window. There's just black. I notice that a couple of people got on at the last stop and have sat themselves near me. I can see the back of an old man's pink-bald dome over there. And there's the tsk-tsk-tsk of leaking teenage earphones from someone else, somewhere near. Two more souls, two more sets of fizzing microtubules, getting closer to up north and physical death with every minute and mile that passes. As we rocket through the hostile, spare depths of the open countryside, I check my watch again. It shouldn't be long now.

Most recently, and partly in preparation for this evening, I've been reading up on some research that's been carried out at the

University of Arizona. Psychology professor Gary E. Schwartz has been experimenting on a medium called Allison Dubois. In one test, Dubois was told to contact the dead husband of a woman she'd never met before – and who was, at the time, sitting thousands of miles away in England. The medium was only given her first name. The information that she then provided about her subject turned out to be seventy-three per cent accurate. But, incredible as this may sound, this was a poor result for Allison. Usually, when Professor Schwartz asks her to 'read' people in lab conditions – that is, people whom the medium cannot see and has never met before – she rarely scores lower than eighty per cent. I wonder, as I pack my bits into my bag, if I'll come across anybody as incredible as Allison Dubois tonight.

I glance to my left to check on our progress. The colours in my window have changed from flat, blank black to a hard, glowing orange as the bridges, bricks and broken fences of the town have emerged from the fields. I pack all my papers away and then listen to what's left of the disquieting testimony of Michele, the reluctant clairvoyant.

'And then I went to visit the family. Well, it turned out that Alan had been to the grave with his wife only three days before with the roses. No one else knew that. Bob's wife did have a very bad cold. And the family rallied round her. She was fine after that. It turned out Jimmy was a nephew who was a very, very good footballer who was about twelve. I don't know very much about what happened to him after that. But certainly on the roses front and the wife, yeah, that was all true. And Graham said, "What was he wearing?" I said, "He was wearing a Pringle jumper." And he said, "That's what we buried him in." I started getting a lot more visitors after that. People just appearing in the bedroom. I didn't even know who they were. I had people talking to me in different languages – I had an old lady who was Greek. I had to try and write things down, but I just couldn't do it. She was chatting to me, standing in the corner of the room. I couldn't always see them clearly, sometimes I could only see their outline. It was freaking me out and I was losing more and more sleep. It was always happening at the same time, just before dawn. And it was happening almost every night. And at the end I was exhausted. I just couldn't work properly, I couldn't think, I couldn't do anything, I was just shattered. And it was getting worse and worse and I was going to bed dreading ... '

And, finally I'm here – the place where, in just over two hours' time, I have an appointment to keep with the dead.

18

'Kangaroo!'

Every Wednesday evening, at half past seven precisely, fifty spectral souls gather themselves together and leave their home in the heavens for earth. They swim down through the clouds, and swoop along the high street, past and around the last few shoppers moseying into Morrison's, over the *Big Issue* seller in the old blue Puffa jacket, and through the tussle of young drinkers who sit and text on the wall by the derelict phone box. They arrive silently, obediently and always on time, at an anonymous brick building in the corner of a Scunthorpe car park to deliver wisdom and messages to a congregation of kindly local Spiritualists. And tonight I will be joining them.

I'd been having trouble getting access to a séance. I'd contacted the Spiritualists' National Union on Maurice Grosse's advice, and when it became clear that I intended to write about what went on, their shrift became suddenly short. I was instructed to submit a formal, written request by post. I didn't expect a reply. And I didn't get one.

Then, help came from an unexpected direction. Big George from Ghosts-UK phoned me up, I mentioned my problem and, in an instant, he solved it for me. Jacqueline Adair, George told me, is a member of the SNU and one of his closest friends. He was sure she would be delighted to take me to a séance.

Jacquie picks me up from the station and drives me back to her small, semi-detached house on the outskirts of town. It's a warm and happy place that glows with welcome. The moment we walk in the door, I'm greeted by a dog, a daughter and a specially prepared roast dinner. Jacquie is the kind of host that humbles you. She's frenetically attentive, checking constantly that I'm not hungry or thirsty or wanting for anything at all. She even insists on giving up her bed for me tonight and sleeping downstairs on her sofa.

I sit at the breakfast bar and, as I prepare to open the first salvos of my attack on the mountain of meal in front of me, I look around. There's a New Age rainstorm of glass baubles, dream-catchers and wind-chimes hanging from the ceiling. The walls are frantic with photos, framed certificates (from the SNU and a Reiki healing organisation) and air-brushed pictures of howling wolves, wondrous starry skyscapes and homoerotic Native Americans on muscular white horses. And there are fairies everywhere. Jacquie collects them. There are swarms of them on shelves, tables and in glass-fronted display cases. A large regiment of the glittery-winged pottery anorexics have even made their way onto the top of her telly.

I put a forkful of beef into my mouth and listen to Jacquie explain some facts about the afterlife.

'When someone dies down here, we cry, yeah?'

'Hmm.' I nod, chewing.

'But in the spirit world they rejoice, because they've come back home. But it goes the other way, too. When a baby is born down here, in the spirit world, they cry.' She cocks her head. 'Do you understand?'

Jacquie is an energetically super-friendly forty-one-year-old mother of four. She wears a baggy, mismatching tracksuit set and a pink plastic Alice-band on her head. As well as being an accomplished Spiritualist, she works from home, during the day, using her powers on a premium-rate psychic phone service.

On the wall, amongst the pictures and certificates, I notice a wonkily drawn pastel portrait of a Native American in a headdress.

'Is that your spirit guide?' I ask.

'Yes, that's Black Elk,' she says.

'And why are they always Native Americans?' I ask.

Jacquie's eyes are suddenly glazed over with blissed-out peacefulness. 'Because they were pure, weren't they?' she says. 'They lived off the land. Do you see what I mean?'

'I think so,' I say

'When you've finished eating, do you want to have a look at my website?' she asks.

'Of course,' I say.

Eventually, when my stomach has surrendered and my plate's been cleaned away, Jacquie sits me in the corner, in front of her PC. She starts to tell me that she used to be a member of Ghosts-UK, but resigned one day in a rage. As she talks, I watch her site, Shanry.com, load up in chunks.

'When you went with G-UK,' she asks me, 'did you come across a guy called Big S?'

'Yes,' I say, remembering the Northern Irishman who led our séance and gave that memorable talk.

Jacquie's eyebrows are scrunched up and her arms are folded. She's taken on the body language of a person making a highly important inquiry. 'Did he follow you all over?' she says.

'Well, yes,' I say, 'but I was in his group.'

Jacquie sighs and a flash of anger whips across her face.

'Yeah, that's just like him. Here … '

She leans over me and interrupts the booting-up of her site to type in a different web address.

'Let me show you something,' she says. I watch the new page appear. It belongs to an American ghosthunter called Troy Taylor. Troy, it seems, is having a conference in June, in Illinois. And his special guest speaker, all the way from the UK, is …

'Oh my God!' I say.

Big S.

I read his biog. ' … Our first international speaker … full-time paranormal investigator … worked with numerous production companies … investigated hundreds of venues throughout the UK … one of the best investigators in Europe … ' And then, I notice his photograph.

'He's holding a cane!' I say.

'That's his prop now,' Jacquie says, a signet-ringed forefinger jabbing at the screen. 'He doesn't do anything without his cane.'

Jacquie's fists are on her hips and her face is criss-crossed with the effort of swallowing the fury. Then she cranks up into a bitter tirade about Big S, her words snapping out and scaring the dog. When she's calmed down, and has given me a guided tour of her voluminous website, she decides she wants to show me something else. She darts off to another room and then returns with a small paperback book. It's called *There Is Always Hope*.

'This is my book,' she tells me. 'It's poetry. Poems about my life.'

'Oh, wow,' I say. I flick it open at a random page. It's a poem called 'Depression'.

There's a small silence. Jacquie is looking at me. I feel a warm puff of embarrassment redden my face. This is too intimate, too soon. I decide to pretend I didn't notice 'Depression'. I glance a look at the next poem.

'Debt.'

She's still watching me. I stare at the page. The dog trots out of the room. I listen to its paws clack on the vinyl floor. It runs up the stairs as I pick another page.

'Divorce.'

The blood in my face runs suddenly hotter. Some windchimes somewhere chime. I flick again.

'I'm Not An Alcoholic.'

Shit.

'Prison.'

No!

'Tramp.'

'This looks great,' I say, closing the book sharply and putting it down on the table next to me. 'Oh, look,' I say as my eyes settle on a serendipitous subject-change opportunity. 'Are those tarot cards?'

Jacquie grins and picks them up. 'Shuffle them,' she says, handing me the pack.

I shuffle the cards, cut them and give the smaller half to Jacquie. She lifts one off the top and lays it down on the table in front of me.

'Ice,' she says. 'And it's facing away from you. That means it's a negative.'

'A negative?' I say.

'Don't worry,' Jacquie says. But she avoids eye contact.

She peels off another card.

'Teamwork,' she says.

'That's facing away, too,' I say.

She turns over another.

'Changing.'

It's facing away.

'Need.'

'And that one is,' I say.

'Gambling.'

And that one.

'Wealth.'

I look between Jacquie and the cards as a small, reptilian panic rears its neck inside me.

'They're all facing away,' I say. 'They're all negative. That's really bad, isn't it?'

Jacquie looks at me and thinks for a beat. Then, she touches the side of her head with one hand and raises the other in the air in front of her, like an aerial.

'Hang on!' she says. 'I've got a connection!' She shakes her head with a cheery, resigned sigh. 'Oh, I knew I'd end up working today.' Then, she fixes me with an intense look. 'March! March is an important month for you!'

'Um … I don't think … no … ' I say.

'A new bed, please! You've just got a new bed.'

I think for a moment. 'No. But my girlfriend *wants* a new bed.'

She smiles. 'Well, that'll be it! Thank you. I've got your grandfather here. He's showing me a watch, please, a watch that won't work. Can you take a watch that won't work?'

Jacquie's hands are in constant motion. They circle each other in frantic, blurring wheels as if they're pulling the knowledge down from the end of an invisible kite-string.

'Can you take a broken watch, please?'

Each one of Jacquie's fingers and thumbs has at least one beefy ring clamped round it. I'm transfixed by them as they circle and move. I scrabble through my mind. I try to think, for Jacquie's sake, for politeness' sake … a watch … a broken watch …

'Nnnno,' I say.

'He's showing me a terracotta wall,' she says. 'Can you take a terracotta wall, please?'

'No.'

Her frenzied sparkling wheels whoosh away.

'What's the Welsh connection?'

'Umm … '

'I'm getting pain. Pains in the stomach area. I'm getting cancer in a female.'

I widen out my eyes and shake my head gently.

'Kangaroo! Why am I being shown a kangaroo, please?'

'Kangaroo?'

'Who's got a speech problem?'

'Um … '

Suddenly, Jacquie clasps at her neck. A look of shock grips her face. 'Huh!' she says. 'Who hung themselves?'

'Nobody,' I say.

Her eyes bulge. 'They had more than one go at it!'

I try to change the subject. 'Um,' I say, 'did you say you had my grandfather?'

'Yes,' she says, 'your father's father. Now, this man didn't say much. But he didn't need to. He just gave you a look and you knew … '

'I didn't really know him,' I say. 'He died … '

'He didn't show much emotion,' she says. 'He didn't suffer fools.'

'Well, that does sound like my *dad*,' I offer.

'Ah, yes,' Jacquie says. 'They were very similar men, he's telling me. That's why he's come to me tonight. He's telling me you were the black sheep of the family ... '

'Yes.'

'And your dad took a lot of your self-esteem away.'

'Yes.'

Jacquie cocks her face and gives me a soft, caring, caressing look. 'This is important,' she says. 'This is why your grandfather has come to me.'

'Right ... '

'He treated your dad the same way. He's telling me he's sorry and that he wants you to know that your dad is proud of you.'

Despite myself, despite the fact that I know exactly what's going on, I feel a marble of emotion suddenly choking up my throat.

'He says that it's going to be OK, but that you need to start taking time out, mentally, for yourself. And you must keep a positive frame of mind. OK? Then you will get your happiness and heart's desire.'

'That's nice,' I say. 'Thanks. Thank you.'

She looks up at the clock near the fairy cabinet. 'We'd better dash,' she says. 'We're going to be late.'

We leave the house, get in the car and begin the short drive to the séance. As we go, I sit quietly for a while and watch the streetlights and shuttered shops of Scunthorpe pass us by. Eventually, I idly mention that I was surprised that it was my long-gone grandad that came through and not my gran, who died in the summer. You see, it was my mother's mother who gave me my first proper experience of bereavement – that is, one that came with the unforgettable phone-call, some tears, a funeral and an awkward post-cremation sandwich event. My grandad died when I was too young to form any real memories of him. But before Jacquie has the chance to properly answer, we pull up outside Scunthorpe's tiny Spiritualist church, just in time for the Wednesday circle.

In the short outer corridor of the church are two doors and a dusty noticeboard. The doors lead to a kitchen and a bathroom, and the noticeboard carries a type-written list of rules, dated 1982. The list bans pregnant women, the under-sixteens and those of an 'unsociable nature' from contacting the dead. It also warns against 'peculiar and hysterical behaviour', including 'extravagant claims of contact with prominent people'. Most of all, though, it forbids 'trance mediumship'. This is the only rule that's been typed entirely in capitals and someone has written it out again, for double-emphasis, in curly handwriting at the bottom of the page. When I read this rule, I get a flash-vision of the roaring man in Coalhouse Fort. I suddenly feel one notch more comfortable about Spiritualism as I make my way into the arena for tonight's main event.

At one end of the room, there's a small stage area with a simple lectern and a number board for hymns. There's also a trestle table with a crêpe-paper-lined box that's filled with donated packets of powdered custard, tinned pears and jelly. In front of all that, there's a circle of plastic chairs with a Spiritualist sitting on each one. There are fifteen of them in all. Mostly, the mediums are middle-aged mothers, but there's also a twenty-something girl with red hair and pins in her nose, and a gangly man with a round moustache and a mournful, determined look about him. He's too tall for his seat and his grey-slacked legs stretch out in front of him, all long and awkward – it's as if someone's forced a cricket to sit in a chair. This is Tony, Jacquie tells me in a force-ten whisper. He is a very senior medium.

After we all trample through a rangy, acapella hymn, a silence overcomes the group. We all avoid looking at each other as we sit still and wait respectfully for the roaming quantum bunches to descend.

After the leader of the circle – a kindly, delicate woman in her sixties with green unsure eyes – receives a message from the deceased husband of a newcomer to the circle, I notice Jacquie's hands revving up in front of her. And then, she says, my gran has come through.

'I've got fish boiling in milk,' Jacquie tells me. 'I can see it plain as anything. And I've not heard of that in years. Years. But I can see that. Now I'm getting the name Iris. Can you take that?'

I shake my head.

'That might be symbolic. Did this lady love flowers?'

'She probably liked flowers,' I say.

'What's the problem with the feet?'

'Um ... '

'Do you suffer in-growing toenails, please?'

'No.'

'I've got Frank Sinatra "My Way".'

'Hmmm ... no ... '

'Might be symbolic. You have to do things your way. OK, I've got a cold. I want pollen or someone who suffers from hayfever, please. It's either asthma or pollen.'

'Farrah has bad asthma.'

'Thank you!' Jacquie chuckles to herself and looks around at all the faces in the group. 'Everybody always thinks that Spirit is talking about them!' she says, to everyone – except Tony, who appears to be, mentally if not physically, in another dimension altogether. He just sits in his seat and stares at his shoes, lost in the universe.

'OK, why would she show me a crucifix, please?'

'Maybe ... she wanted me to go to church more?'

'OK, thank you. Would she think we were, I can't think of the right word, like, devil-worshippers?'

'She might have found all this a bit strange.'

Jacquie nods firmly. 'Right, yes. Thank you. This is what I'm being told. Right, there's farming.'

'Farming?'

'Cats,' says Jacquie.

'No.'

'And trains. Something to do with trains. A signal box.'

'I'll have to check that out ... '

'Why do I have three sons, please?'

'She had two sons.'

Jacquie frowns to herself and concentrates. We watch her in silence. All of us except Tony, who's still staring sadly at a space in the middle of the floor. The group's leader, I notice, keeps shooting the senior Spiritualist concerned glances.

Jacquie looks at me again. 'I've just asked if everything I'm getting is correct,' she says, 'and even though it's scattery, she's told me definitely three sons.'

I sit up and forward. 'Hang on,' I say. 'You're right. She had a baby boy called Richard who died.'

'And the number three,' she says, 'that's important to you.'

'Well, someone told me … when the number three crops up I should be really careful.'

'What about when you fell off your bike and cut your chin?'

'No, that hasn't happened … ' I say.

'And your nose was broke.'

'No … '

'What you cannot take will come around,' she says.

'Oh.'

There's a quiet pause as I let that sink in. And then, Jacquie's off again.

'She don't like Christmas.'

'Didn't she?'

'She was quite abrupt. She doesn't like you slouching. She was a very strong-willed person. She didn't take any nonsense.'

'Yes, yes …'

'Did you not get to say some things that you wanted to say?'

'Well, no … '

'She knows what you wanted to say. She says you can't carry the guilt for that. She's telling me she's fine. To be honest, I think it's right that she went when she did because a lot of things were slowing her down.'

'That's right,' I say. 'She'd just had a prang in her car. They were thinking of taking her driving licence off her.'

'And it was a good way for her to go. She was fine with it.'

My gran's greatest fear was that the creeping rot of age would take her mind, her dignity and then her home. But as it was, she died, as independent and as sharp as she'd ever been, in her favourite chair in her living room, with a freshly made cup of tea by her side. She just closed her eyes and her microtubules emptied. She was ninety-two years old. Everybody agreed, just as Jacquie has said, that it was a good way for her to go. Perfect, in fact.

'That's absolutely right,' I say.

Jacquie's hands slow rapidly. 'OK, thank you,' she says. 'Now you've got your proof.'

I sit back and wait quietly for the hour to be up. And when it is, the older members all turn their attention to Tony.

'Not get anything tonight?' Jacquie says to him.

He's looking at his knees, exhaustedly. 'I've been doing something,' he says. 'They're in crisis up in the spirit world. They're in a state of great distress. Because of the tsunami. They're flat out. We usually have fifty guides with us in circle and tonight, we only had eight. So I've been sending energy up to them. You can help if you want. Imagine you're making a balloon in the middle of the floor out of energy, then send it up. Can we all do that for five minutes?' He looks around the group with wide, pleading eyes. 'Every little helps.'

So there we sit, fifteen Spiritualists and me, in a tight circle, in a cold brick building in the corner of a Scunthorpe car park. And, silently and selflessly, we all do our bit for the tsunami victims.

19

'I talk to the devil every day'

So my journey has led me here, to the centre of the web, to the teeming heart of the Catholic empire. Somehow, it seems inevitable. I have come to the Vatican, the capital of everything I raged against at school, the centre of what I've always thought of as the world's biggest and most powerful supernatural movement. I've spent the morning walking the ancient, grand and immaculate rat-run of streets and domes. I've looked up at the old gold ceilings, I've hurried past the opulent palaces and I've seen the holy treasures they display in this beautiful and sacred citadel, the wealthiest country on the planet, and the one that's closest to heaven. And now, after everything I've seen and read over the months, I don't know what to think.

I've come to rest in St Peter's Square, amongst the hawkers and the tourists and the nuns. I check my watch and stand up, as a cold early-winter wind tumbles down through the square and ruffles the coats of a squadron of strutting pigeons. Grey feathers on grey ground under a great grey darkening sky. I zip up my coat and hurry to my appointment with Father Gabriele Amorth at his base, deep in the mix of streets in a southern suburb of Rome.

Amorth greets me quietly. His face is lined with long years and weariness. He wears a black-buttoned tunic, has large, lobey ears, and is dignified and peaceful and firm. He leads me into the complex of chapels and dark, hushed passages. I'm struck by the tranquil sense that holds this place. It's as if it exists in a cocoon, a holy bubble of serenity. Then, suddenly, as we walk down one particular corridor, the feeling changes. Amorth takes me to the end of the uneasy hallway, opens an unmarked door and shows me into a cramped and shabby room. I pause and look around. So this is where it happens.

The paint on the walls is aged with the dirt of years and there's a deep, black crack that runs upwards through the plaster and spreads out towards the ceiling like the fingers of a skeletal hand. Somebody has pinned up pictures of long-dead Catholic holy men. They gaze out, with sorrowful eyes and thin, dry smiles, into the heavy air that hangs in the room. They see nothing, but they say everything, as they stare straight through the gloomy objects that fill this place – the bed and the chair and the statue of Mary and the handmade wooden box that's been nailed in the corner at about waist height. It holds the restraints. Tough old rags that used to be white but are now dark with the sweat of the people who have been taken here and tied down.

'They are used for people who are furious,' says the exorcist. 'They are used for tying arms and legs. They lie here,' he says, pointing.

It's like an old hospital bed. It's high, with tubular metal legs, and its head is raised up. The faded blue cloth that covers the mattress is stained in the areas where hundreds of heads and bodies have lain down and struggled against the invading magic as Father Amorth has murmured his ancient Latin prayers over them.

'And here –' He turns towards a corner table that carries a cluttered huddle of bottles and pots '– is the holy water. And here is the blessed oil and here is the blessed salt.' He takes a step out of the room and opens another old door. I follow him out, then the eighty-year-old cleric stops and turns to look at me. His eyes have taken on a solemn and portentous drawn gaze.

'And this is the bathroom,' he says.

'Oh right, thanks,' I say, before following him into an anteroom to sit down.

I'd all but given up trying to find an exorcist to speak to. First, there was the American pastor who reacted to my simple request so memorably: '*You are playing with fire.*' Then there was Stan's local churchman, who panicked on the phone and – I'm sure – was behind the abrupt cancellation of my interview with him. Next I tried the Catholic Church's press office and they arranged for me to meet a priest in Cardiff. His first words to me were: 'Me? An exorcist? I'm sorry, I think you've been misinformed.' He went on to tell me it was his firm belief that science held all the answers. I'd been stitched up, palmed off with a P.R.-conscious progressive. At a loss, I asked my mother for help. I thought if an insider made the approach I might get somewhere. And she tried. But even my fundamentalist Catholic mum was told, in no uncertain terms, that there would be no interview. Some months later I was informed by a senior churchman that there are only two full-strength Catholic exorcists in the country, and their identities are kept strictly secret. Every parish, I was told, does have a priest who can perform various minor 'deliverance' blessings and this, I realised, was the position Father Bill must have held.

Lou Gentile was right. The twenty-first-century Church doesn't like discussing its demons in public. But before I gave up completely, I decided to fling my hopes to the wind to see how far they'd fly: I took my request right to the top. Father Gabriele Amorth is the Vatican's chief exorcist. He's the papal commander-in-chief in the war against the devil. As Amorth's name is in the public domain, I thought I might have the slimmest chance of a chat. And, to my total astonishment, he agreed to meet me.

I sit down at a small table in front of Father Amorth and tell him about the science priest that I met in Cardiff. He listens stilly and, when I've finished, he lowers his head to look at me.

'It is true,' he says. 'Some clergymen do not believe in the devil's activity. This is because they have never studied exorcism, they have never carried them out. It is a problem. I questioned the Pope about this and I will tell you the answer that he gave me. He told me that if they don't believe in the devil, they don't believe in the scriptures. This is Satan's work. The devil uses all his resources to stay hidden. He tries not to be believed in.'

Amorth talks about a sceptical priest who attended an exorcism with his arms crossed sulkily and a sarcastic grin smeared across his face. He remained like that, watching, until the possessed child turned to him and said, 'You say that you do not believe I exist. But you believe in women! And how!' The womanising priest, horrified, humiliated and thoroughly rumbled, walked backwards out of the door and fled. This is a perfect example of an apparently possessed person displaying knowledge of something they shouldn't know. And it's astonishingly common in Amorth's patients.

'I always pray in Latin,' he tells me, 'and even though the possessed person doesn't understand Latin, they understand the prayer perfectly. Even the young children. And if I make a mistake, they laugh and make fun of me. At that moment, it is the devil who is laughing, not the child. I use Latin because it helps me understand if they are possessed and also because it is the original language used and it is more powerful. The prayers are extremely old. There were made official in 1614 and taken from centuries before, for example, the exorcism of Santa Rosa in the fourth century. Now there are new prayers that have come out, but I don't like them. They don't work.'

I was about to ask Father Amorth how he can tell the difference between a 'genuinely' possessed person and someone who is mentally ill. I suppose that a youngster suddenly acquiring a working knowledge of Latin and ancient Catholic rites would certainly trigger suspicion. And I'd also note that the victim is mocking the priest, just like Kathy and the others. When I do ask the question, however, Amorth acknowledges that it's often difficult to tell the diabolics from the daft.

'It is true,' he says, 'it is a problem deciding which person is possessed and which needs a psychiatrist. My first exorcisms were false – I didn't understand that they had psychiatric problems. It takes a lot of work. But I'll give you an example. A father thought that one of his sons was possessed by the devil. At the table while they all ate together, the father said, in his mind, "Ave Maria." It was only in his head, but the son got up very quickly and said, "Enough! Stop it, Father!" This is a phenomenon that cannot come from psychiatric illness. There are also other ways I can tell. They will have an aversion to the sacrament and all things sacred. They will scream, spit, or vomit. They will go into a trance during the exorcism and often become furious. They can become very strong. During one exorcism, I saw a child of eleven held down by four strong men. The child threw the men aside with ease. Often I keep my fingers on the eyes because when the devil is there the eyes go very high or very low and you do not see the colour, just white. Sometimes, strange objects come out of the body. I've had cases where they spat out nails, plastic animals, keys, chains. And sometimes they levitate. A twenty-five- or twenty-six-year-old farmer was sat on a chair and held down by eight people. He was very furious. During the exorcism he levitated to the height of two spans of the hand. This happened various times.'

'The levitation you describe,' I ask. 'Is it common?'

'It is not common,' he says. 'But it does happen. I will give you another example. It was an episode in Africa during an exorcism of a young person in a church. The doors were closed and the only people present were the family and a few other people. During the exorcism, this person began to rise up higher and higher until her head touched the ceiling. This was a very diabolical episode.'

'Have you ever been hurt?' I ask.

Father Amorth swallows and looks down at his plump, pink and ringless fingers. 'Only once,' he says. 'The possessed person was on the bed and he moved his leg. It only seemed like a small movement, but it broke my leg.' He looks up. 'They have huge strength.'

The exorcist speaks slowly, without emphasis or passion. It's as if he's answering my questions through a sense of duty. He has only the faintest hint of grey hair remaining on the sides of his head. It's so wispy that it looks as though there is an out-of-focus patch of air or a thin mist above each ear. The smooth, flat shine on the top of his head contrasts with the loose slips of flesh that hang down from his eyes, lips and neck. It looks as if his skin has been poured onto the top of his skull and is dripping down off his face, slowly. Being with Father Amorth and listening to him speak, I feel humbled, fascinated and compelled to believe every word. It's his steady delivery, the steely serenity of his eyes, the dignity and composure of his black robes and bearing. The steam that's coming off the exorcist is learned, powerful and gripping, and I suddenly find that I'm feeling at home amongst the crosses and the incense and the prayers. *Oh, you will be.*

'What I don't understand about possession is,' I say, 'what does Satan get out of it?'

'He wants to show that he is more powerful than God. It is a constant battle. He says that by possessing somebody it shows his power because God doesn't have the power to send him away.'

'And have you seen the devil?'

'No. I've never seen him. I don't need to. I talk to the devil every day. I have touched the invisible world with my own hand. Besides, the devil is a pure spirit. It's not a physical form.'

'What do you say to him?

'Always the same things. I ask, "Are you alone or are there other demons? What is your name?"'

'Do the demons have names?'

'There are many names,' he says. 'Zebulun, Meridian, Asmodeus … '

'Have you heard of Hecate?'

There's a silence. Amorth looks at me with his sunken, wet eyes. 'No.'

In an echo of what happened after my first meeting with Maurice Grosse, a die-hard rational part of me is finding Father Amorth's testimony too powerful, too extreme. *It just can't be true.* But, really, I counter to myself, to call the exorcist's integrity into question is just a cop-out. Why would he lie? Why would he do it?

To suggest that this exceedingly senior churchman has simply invented all these tales of super-human toddlers, Latin-fluent farmhands and floating, furious Africans is just unreasonable. It's an excuse, isn't it? Because the Blockage doesn't want to believe.

Well, I've got some things to say to the Blockage. For a start, what Amorth is telling me is absolutely consistent with what exorcists have reported for centuries. Then there's the small movement of a possessed person's leg that led to the fracture of Amorth's bone. How can that be explained? That's nobody's imagination or super-aroused hallucination. That's a plaster-cast for twelve weeks. And I'm equally intrigued by his claims that the new exorcism prayers don't work. In 1999 the Vatican updated them and Father Amorth has since called their new version 'a masterpiece of incompetence' because he's found them to be ineffectual. So, my question to the Blockage is this: if the effects of his exorcisms are *really* all in the minds of the priest and his patients, how come one prayer works and another one doesn't?

And Gabriele Amorth is no simple hick. He was born in Modena in 1925 to a family of lawyers and judges. During the war, he joined the Italian resistance. And as soon as Hitler did the decent thing and Amorth was released from his freedom-fighting duties, he became a member of the newly formed Christian Democratic Party. He was the deputy to Giulo Andreotti, who became prime minister seven times over. In short, the priest who's sat in front of me is no sceptical monsterologist. He's no trance medium. And almost everything he's saying is consistent with what Lou Gentile told me all those months ago about demons, exorcism and the many dangers of divination. With that in mind, I decide to fill the priest in on the Enfield case, in particular the fact that the unsettling north London narrative has a Ouija board and an exorcism as its prologue and epilogue. I tell him Janet told me that this was a 'coincidence'.

Amorth responds with an almost imperceptible shake of the head. 'No, I don't think so,' he says. 'That was not a coincidence. The Ouija board is dangerous because a person can become possessed by the devil. They call the souls of the dead but the dead souls never come. The dead don't move. But the demon can come.'

I tell him about the old, mocking male voice that Janet would speak in, and that it was in possession of surprising facts about the house's previous occupant, Bill ...

'It was the devil,' says Amorth. 'It was the devil who made her speak and say these things. It is rare for something like this to happen.'

... and about the unforgettable subterranean gaze that she suddenly took on and that it was the same as Kathy's ...

'This would happen with someone under a very strong possession,' says the priest. 'The force of the devil can scare people like this.'

... and about the bangings and the crashes that are common to so many of these cases.

'It is possible for a place to become possessed,' he says. 'For example, the windows and doors open and close without anyone touching them. You hear lots of bangs. The television turns itself on and off, the lights turn on and off without anyone touching them. There have even been flying chairs and dancing armoires. I've had to do various exorcisms of houses. To free a house takes a long time. But less time than a person. When freeing the houses you use holy water and salt. You put the salt in the corners of the room. You also put salt on the windows to avoid an infestation. You use the salt against evil presences. The salt protects a place. I also use oil and water. I put the oil on the eyes, ears, nose, mouth and throat. All the senses. With the water I bless them at the beginning and during the exorcism if they become furious. I bless them again and they become even more furious.'

I think about the last haunted place I stayed in – the First And Last Inn in Sennen. I'm still troubled by the behaviour of the pets, especially the moment when they stared at Annie's wall and then bolted. Could this, I ask, be the sort of thing that happens when a place is possessed?

'Yes,' he tells me. 'Animals are sensitive. They usually fix their eyes on one point, they stay there and then they run off. They are afraid of whatever they are looking at or feeling. Animals are very interesting regarding their behaviour. I get a lot from them.'

Amorth says it can take years for a deliverance to be successful. He tells me that other symptoms of possession include vicious pains in the stomach and head that doctors can't diagnose. The agony often gets so bad that the victims attempt suicide. Amorth says that many doctors and psychologists refer their patients to him when they sniff Satan in a case. He describes an apparently epileptic boy who was cured with an exorcism.

'That boy wasn't ill,' Amorth recounts. 'He was demonised. The devil does this to hide himself. He gives the person symptoms of an illness. This way people believe that they are ill when there is actually possession by the devil. The doctors can't find any cure. For them, the medicines don't work. After his exorcism, the child was completely healthy.'

But things don't always end so cheerily. In 1978, in the Bavarian town of Werneck, Father Ernst Alt and Father Wilhelm Renz were found guilty of negligent homicide after the exorcism of a twenty-three-year-old girl called Anneliese. Eleven months before her death, her parents – who, along with the priests, received a six-month suspended sentence – turned to the Church after becoming dissatisfied with the medical establishment's diagnosis of her daughter as an epileptic. Renz and Alt carried out sixty exorcisms over a period of nine months. The sessions were often violent, and the demons in Anneliese would pretend to be Adolf Hitler, Judas Iscariot and Emperor Nero. From the start of her ordeal, the pretty, dark-haired theology student refused all food, drink and medical attention. She died, on 1 July 1976, of malnutrition and dehydration. She weighed slightly more than three kilograms.

'Yes,' says Father Amorth. 'I know this case very well. Anneliese Michel. She was a girl who was offered to God for the forgiveness of our sins. She died for God. God allowed her to be possessed by the devil. They blamed the exorcists because she died whilst she was being exorcised. Then they prosecuted the exorcists and the girl's parents. The prosecution was ridiculous and unworthy. It wasn't the fault of the exorcists or the parents. It was the will of God.'

'What I don't understand,' I say, 'is if it was the will of God for her to suffer and die, how can we then say that He's merciful and loving?'

The exorcist doesn't blink. He continues, slow, still, unmoved. 'He permitted the death of his son Jesus,' he says. 'He permitted the death of martyrs and lots of people suffering from illnesses.'

'But surely that's the behaviour of something evil?'

There's a freighted pause. 'No,' says Amorth. 'He is good because there is a need for this sufferance to make saints.'

'So we need people to suffer so they can become saints?'

'Yes.'

'Why do we need saints?'

'To protect humankind.'

'But what good does their suffering do to the rest of us?' I say. 'I don't understand how we can benefit from these people suffering.'

Amorth's eyes peer up at me. From underneath his ageing folds, he looks like a slow, basking reptile who's had his afternoon dose disturbed by a titchy, skittering gnat. 'Because it gives us the opportunity to repent our sins,' he says. 'The sinners abandon their sins. They get a better life. And people convert. Anneliese's death absolved the sins of others and it converted many people.'

'So a lot of people converted to Christianity because of this?' I ask.

'Yes.'

'And this means that Anneliese is a saint?'

'Yes, I think she is a saint.'

I look at Father Amorth. Father Amorth looks at me. There's a silence. I look down at my notes and fiddle with the corner of a piece of paper. I decide to change the subject.

I've been thinking, these last few days, about the things I've been up to on my journey – the Ouija boards, the séances, the demonically infested houses, the graveyards, the cups of tea I've drunk that have been prepared by possessed women. And I've been wondering what Father Bill, if he were still alive, would have to say about it all. And, with all that in mind, and while I'm here, I think, well ... I might as well ask.

Father Amorth reacts to my request for an exorcism with typical steady understatement. And, thankfully, he doesn't tie me down. He stands at my side, rests the end of his purple stole on my shoulder and begins his prayer. '*Ecce crucem Domini*,' he begins in a low, dark hum. He dabs the blessed oil on my head, eyes and lips and rests his hand on my head. '*Sit nominis ti signo famkus tuus munitus*,' he murmurs. I sit back and wait for the eruption. I feel tense and sweaty and primed. '*Dei exorcisto*,' he says. And I don't spit or scream or vomit up a nail. I'm not furious. I actually feel quite nice. A warm, relaxed swell emerges in my ankles, fills up my legs and spreads through my torso and arms. I let go a sigh and decide to relax into my exorcism as if it's a big, soft, spiritual sofa.

'You are not possessed by the devil!' announces the exorcist with a big smile when he's finished.

'Excellent!' I say, and leap to my feet. I follow Father Amorth out of the anteroom and into the car park, which sits in the middle of the complex in the classic Roman atrium fashion. The moment that I get outside I'm struck, yet again, by the potent feeling that permeates this place. I felt it when I walked in. There's something in the air out here. It's as powerful as the feeling of dread that I felt in that upstairs room at Michelham. And yet, it's the precise opposite. I feel as though I'm sinking slowly into a warm marshmallow of love. I want to pitch up a deckchair, unbutton my shirt and just bask, for a while, in the soothing magic. I can't understand where it comes from. It can't be Dr Salter's 'psychic *feng shui*' as this is just a standard car park. There's worn tarmac and unremarkable vehicles and an awkward family of streetlights. And, backing onto it, there's the high concrete wall of a triple-decker church. Could it be the powerful concentration of holy worship in there that's leaking into the air out here? This sensation, I think as we walk, is the same thing as the 'atmospheres' that people talk about. It's like when Janet said 'there was a funny atmosphere in the house' on the night that the terror first kicked off. And when home-buyers are shown around houses on property programmes, they're forever remarking that 'it feels nice in here' or the opposite. Is this paranormal? Is it the same thing that Stephen the Druid taught me about – a sixth sense?

We walk back inside, and down a long corridor whose dusty silence is broken only by the exorcist's steady talking.

'People don't realise the power of the devil,' Amorth says. 'And if we don't realise his power, we can't fight against it, and he continues to do what he wants. There are not lots of people who are possessed but there are many people who follow the devil. Because the devil has two activities. One ordinary and one extraordinary. The ordinary activity is temptation. Tempting man to evil. The extraordinary activity is possession.'

'So big world issues,' I say, 'like Iraq. Is this the work of the devil, tempting Blair and Bush?'

'Yes, because God is the God of life. The devil searches for death. All wars are the work of the devil.'

We enter a meditative high room and, eventually, reach the front door. The moment I open it, the outside crashes in. It feels like a different world out here. The evening is bearing down, by now, over the spires and the traffic and the people. There's bustle, shops and tinny Latino pop coming from a flat over there. There are planes in the sky and peeling, graphitised billboards and a special meal deal on at the grimy local McDonald's round the corner. Traffic lights change and ping at the citizens and they march across, purposeful and pre-programmed, each with a sprinkle of the universe buzzing deep in their brain-stems, each absorbed in their own invisible worlds. I stand there and look around at this world of things, where the devil and his enemy have all but disappeared from sight. And it *does* feel like a different world out here. But is it?

'Father Amorth,' I say, 'is the devil winning?'

He blinks slowly and nods. 'Yes. Our world has many diabolical influences and there are many temptations. The devil is strong. He is winning. God leaves us alone. He lets us do as we do. But I believe that there will be a very big punishment of the world and then we will return to the path of God. I believe in a Third World War.'

20

'I am the ghost'

The dining table in the most haunted house in Britain is protected by a thick, rubbery sheet. Covered in concentric diamond patterns and the colour of cheaply recycled paper, its surface is strafed with small, brown stains. These are the tablecloth's war medals, proud evidence that it's taken the hot beverage bullets many times over for the antique wood that it protects. I put my tea mug down on it, as I sit and start to talk to my host. But when I look up, he's not there. Before I have the chance to locate him, he darts back into view and slips a wooden coaster underneath my mug. He is wearing a wounded look that morphs, slowly, into a minor scowl.

'Oh, sorry,' I say. 'I thought … '

But I trail off. It's clear by his reaction that the Founder doesn't want to talk about it. He looks hurt, personally hurt. It's as if I've slammed my steaming drink down on the back of his hand. There's a tense silence. The Founder sits down in front of me and puts his own cup on a coaster. I smile at him and try to break the gooey awkwardness that's swollen up in the air between us.

'So, do you get more frightened here than you do anywhere else?' I ask him.

'Well, I wouldn't say frightened,' he says. 'I think a better word is … um … what's a better word? Daunted.' He takes a careful sip of his tea, bending his head to meet his cup. 'You have to be careful what you say,' he says. 'Saying, "I get frightened" puts you into a different spotlight. You get people saying, "He's a ghosthunter and he gets frightened?" You know, I've got thirty-six years' experience doing this. I'm in the league of your Andrew Greens and your Maurice Grosses.' He puts his mug down, slowly. 'It can get frightening for other people. But I love this house. I absolutely love coming here. It's where I'm happiest.' He looks around for a moment. 'This place is like a spider's web,' he says. 'You get drawn in.'

I can't tell you where I am, because I'm not supposed to be here. But I can say this much: the thirteenth-century property in which I'm drinking tea is internationally legendary as being the most haunted place in Britain. It was a 'mass house' during the Reformation and secret Catholic services were often held here. Illegal clergy used the building as a hideout during those murderous years and the walls of this place are pocked with priest holes and drenched in evil history.

For a while, paranormal enthusiasts were allowed to come to spend the night, for a small fee or, most often, for charity – and in the cabinet behind me there are several thick folders filled with hundreds of event logs that attest to this place's dark madness. More recently, though, the property has been bought privately, and its new owner is fed up with the regular annoyance of people knocking on the door asking for tours. So, he doesn't want publicity. He's not even been told that I've come to stay for the best part of a week. Currently, he's far away on holiday and David Vee, founder of Ghosts-UK, is house-sitting, as he does whenever the building would otherwise be unoccupied. Maurice Grosse told me that I should stay in a haunted location for a length of time to get the best spectral results, so I was thrilled when the Founder said I could join him for a few days here. I couldn't have hoped for a better, more dreadful place than this.

'I can prove that ghosts exist just by the stuff I've got from this house,' the Founder tells me, his eyes looking large and strangely vulnerable through his thin wire glasses. 'But I'm not allowed to use it. Because the owner won't let me. That's one of the strict instructions. I'm a friend of this house. I have a bond with it and the owner, and I wouldn't do anything to jeopardise that bond. Plus he'll sue the arse off me.'

Outside, the fields are bitter and lifeless and the trees are petrified and skeletal. I take a sip of my tea and put it carefully down on the coaster that David has placed on top of the rubber tablecloth. I'm very careful not to drip. At this stage, I just think that he is being slightly over-zealous in his house-sitting duty-of-care. I soon come to realise, though, that the Founder's relationship with the most haunted house in Britain is far stranger than that. It's the way he talks about it obsessively, the way he looks at the walls and handles the furniture, the way a touch of the love rival's jealousy rears up in his voice when he mentions the owner and other visitors that have been allowed to stay here. I'm not sure what's behind this curious and oddly moving behaviour, yet. I'm only just aware of it. But the Founder loves this place *too much.* He haunts it.

' ... it's only true poltergeist activity if it moves in an arc of parabola,' the Founder is saying.

'Sorry, what?'

'I was telling the team from *Haunted Homes*, this show that I've been filming for ITV1. When an object moves of its own accord, you can only confirm that it's true poltergeist activity if the object moves in an arc of parabola.'

'What's an arc of parabola?'

'When I say that, people are always extremely impressed, but all it means is a semi-circle. If you call it an arc of parabola, it makes you sound more important. It's the same with words like "parabolic", "elaborate", "paranormal", "pertaining to".' The Founder looks at me and smiles with just one side of his mouth, before saying to himself, a little wistfully, 'Arc of parabola.'

I shift in my seat, take another sip of tea and try to think of something to say. Suddenly, I realise that my mind is completely empty.

'You've lost weight,' I say.

'Yes, I have,' says the Founder. 'I've lost about seven stone.'

'*Seven* stone?'

'Yes.'

'Wow.'

I take another sip of tea, and rub my fingers over the surface of the tablecloth for a bit.

'Seven stone!' I say and smile at him. I put my cup down in the dead centre of the coaster.

There's a silence.

'It's quiet,' I say.

David listens for a beat. 'It's eerily quiet,' he says. 'The problem with ghosts is they check you out. They're intelligent.'

There's another silence.

'Could they be checking *me* out?' I say.

'I think that's quite possible,' he says. 'They could be waiting to pounce.'

I look around me, at the pictures and the windows and into the corners. 'How many ghosts are here?'

'We've had sightings of thirteen different entities,' he says, folding his arms in front of him on the table.

'Can I have a look around, then?' I ask, getting up from my seat. I'm anxious to have a wander about. There are two rooms upstairs that are especially notorious. 'Slow down!' David says. 'Bloody hell! Don't be in such a hurry all the time. You're not in London now, you know.'

I sit back down and help myself to one of the folders of ghost reports. I flick through the pages. After an hour or so reading, patterns begin to emerge. As well as the usual footsteps and self-closing doors, many people have felt an invisible hand holding theirs, had their hair pulled or felt suddenly overcome with a powerful feeling of sadness in the child's bedroom, upstairs. But the room next to that is said to be the most troubled of all. I read an interview with the house's previous owner that's been cut out of an old local newspaper. On her first night here, the old lady and her husband prepared for sleep with the door closed. As soon as the light was switched off, the latch popped up and the door swung open again. This happened four times. After the fifth attempt at closing the door, she commented to her husband, 'If it does that again, I'm sleeping downstairs.' It did it again. And a full apparition of a priest appeared in the doorway. Despite the fact that it's in the most haunted part of the most haunted house in Britain, this room is still referred to as 'the haunted room'. And it has remained unoccupied for decades.

David is still sitting opposite me. He's smoking another cigarette. I keep catching him glancing at me. I smile at him through the silence.

'When we were in Wales,' I say, 'you told me that you were attacked by three apparitions in this place.'

'Yes,' he says. 'That was extremely odd. I was on my way back from the kitchen and I saw a reflection in the back of my glasses. I turned around and there were three separate entities, full apparitions, as solid as you and I.'

'Fuck!' I say.

'Absolutely. They were all dressed in different periods. One was from the fifteenth century, one was from the sixteenth century and one was from the seventeenth century. It got quite daunting. I wouldn't say frightening because I don't get frightened.'

'What did you do?'

'I suppose I cowered by the door, thinking, oh my God, they're going to attack me. They started to approach me quite quickly. But it was as if they couldn't see me. They could hear me, but they were looking right through me.'

'Almost as if you were the ghost,' I say.

He considers for a moment, then takes a drag of his cigarette, nods and blows the smoke out. 'Yes, that's absolutely right,' he says.

David looks a lot older since I last saw him. He's got bleached highlights in his hair. As his smoke drifts upwards and leaks between the gaps in the floorboards of the haunted room upstairs, he looks into the middle distance, suddenly in the clutches of another, secret thought. I hear him murmur, 'I am the ghost.'

An hour or so later, we have unexpected company. A young friend of the owner's has also decided to take advantage of his absence, and has invited his girlfriend and a clucking gaggle of her mates around for a mini-vigil. They've brought cans of lager, vodka, Coke and a lappy, hyperactive blond dog called Charlie. As they sit around the table, with their drinks not on coasters, the Founder flits between eyeing their crisp crumbs, fag ash and miscellaneous detritus crimes, and stunning them all into a fish-mouthed silence with his special expertise. They're awestruck by David's tales. The women have communal lipgloss, huge hoopy earrings and hair the colour of chips. As they lean forwards on the table and listen to the Founder, I sit back and find myself confronted with a triumphant parade of frilly thongs flying inches above the waists of their jeans.

'You can point your camera anywhere you like in this house,' says David, 'and I guarantee you'll get an orbicular light anomaly.'

Everyone looks extremely impressed.

'Do you want to hear the famous scream tape?'

He's beginning to enjoy himself now. He walks over to an eighties tape player, which is the size and shape of an upright briefcase, and pushes down the play button, heftily. The girls sit forward in silence, as the smoke from their Bensons gathers into a large milling cloud over the table. This recording was made a decade ago. It is chilling. The first thing you hear is a sole, agonised scream. That's followed, instantaneously, by louder group screaming that sounds like it's borne of a confused and panicked terror. The first scream, David tells us, was heard by some charity ghosthunters at between five and six in the morning. The second is their reaction to it. Like the Enfield tapes, there's an unmistakeably authentic, non-*Crimewatch*-reconstruction edge to the recording that has all of us making worried eye contact around the table. For a silent moment, everybody tries to work out what everyone else is thinking, before we all break off into a murmuring clamour of 'bloody hells' and 'oh my Gods'.

The house's owner has apparently had the tape analysed by

specialists at Manchester University. Not only have they confirmed that the first sound is not of human or animal origin, but they also concluded that it actually seems to be made up of five separate screams.

David then plays us some recordings that he's made here during the night as he's slept. There are footsteps, metal door latches being popped open, doors slamming and, most fascinatingly, a recording he made in the upstairs room where the workings of the drawbridge used to be. The tape sounds, incredibly, like a drawbridge being slowly raised. The most bizarre recording in his collection, though, lasts for about ten minutes and it sounds just like the swishing, clashing foils of a fevered sword fight.

'You see, I think,' David says to me, taking the tape out of the player 'in our time and their time, the difference, although only a matter of minutes to us, is a lifetime to them. Take one of the recordings I got from this place. It's five minutes long. But it only took five minutes to get here from the fifteenth century. They transmitted it in a five-minute segment and it took eight hours to receive it.'

I have no idea what he's talking about.

'So ... where did the sound start?' I say, blindly feeling for an entry point. 'Medieval times?'

'In the fifteenth century,' he says. 'And it took eight hours for me to actually receive it'.

'How do you know?' I ask, fumbling haphazardly.

'Because it came from the same location. It came from here, so it should be here all the time. But it wasn't. It travelled. It came back. And it was pitched up by seven semi-tones. So then I had to slow it down seven semi-tones because it got here too quickly.'

'So,' I say, 'when you say eight hours ... where's it been in that eight hours? It started off here ... '

'It started off here and, obviously, because of the frequencies, it then takes eight hours to work its way back again. And the reason I knew that is because the electro-magnetic frequency meter-reading was constantly on fifty milligaus, which is an extremely paranormal phenomenon.'

There's a long pause. My mouth is a bit open.

Suddenly, thankfully, one of the girls interrupts our confusing stand-off.

'Why did you start doing this, then, Dave?' she says. 'What got you into it?'

'It all started when, as a child,' he says, 'I had a ghost living under my bed. It had a cloak. It was quite terrifying. I kept telling my mum and she kept saying, "Don't be silly, ghosts don't exist." And a couple of months later, she walked in just as the ghost appeared out from under the bed. The expletives were unbelievable. Soon after that we relocated.'

'You moved?' she says, with crisps.

'We moved house, yes. Does anybody want to come upstairs?'

We all stand and follow David up the stairs into the child's bedroom. There's a standard lamp, a couple of old armchairs and a thin rug over aged bumpy floorboards. There's also a priest hole with its hatch off, low on the wall. Just like the 'haunted room' next door, this place is unused by the owner and the air inside it feels cold and angrily forbidding.

'Is it true that there's a head buried in the grounds somewhere around here?' I ask.

'There's no circumstantial evidence for that,' David says, 'so we don't know.'

Behind me, I become aware of a small commotion. I turn to look. The larger of the guests has sat herself down in one of the armchairs. Her face is in her hands. Something is wrong.

'You all right, Sam?' says one of the girls.

'I just suddenly feel really ... ' She breaks off and sniffs. 'I just feel really, really sad,' she says, pushing a weak smile out through the misery.

I decide not to say anything, but just make eye contact with David. It's been obvious since their arrival that these people aren't familiar with the details of the phenomena in this house.

I wonder if she could just be reacting in an emotional way to fear, to the adrenaline that might be flooding her blood. But why in this room? When it's here – and only here – that I read several reports of the same thing happening, years ago? Just as I'm thinking this, and over the top of the group's bubbling chatter, there's a loud crash downstairs. It sounds as if someone has slammed a chair against the wall. Charlie the dog, who's still down there, starts barking urgently. Everyone's absolutely still: a mini, momentary group petrification grips us. Then, the girl in the armchair looks up, her eyes red and wide.

'What the fuck was that?' she says.

'Shit,' I say.

'Shall we go and find out?' says David.

But there's nothing down there – just the room, exactly as we'd left it. Bags and crisp packets and ashtrays and the boiler, chuffing lazily to itself.

An hour later, the group of friends has got bored and gone home. And the Founder is not happy. He's furious about the presence of Charlie the dog. He scrubs the stone floor where it knocked over some water, puts all their rubbish in the dustbin outside and Mr Sheens a chair where a girl with perfume was sitting.

'I hate that smell,' he says as he rubs, 'it makes me gag. Ur! Disgusting. Ur! And that bloody dog was totally out of order. I do get annoyed when people abuse this place. But I have to bite my tongue because I can't say anything. Apart from if they do something terribly wrong upstairs. Like if they take a drink up there. Hang on,' he says, pausing with his duster and his can, 'I can hear sounds, like wooshing sounds.' His ear is cocked skywards. 'It's almost like windy conditions, even though it's not windy outside. Can you hear it?'

'Yes,' I say. 'Is it an aeroplane?'

'No, I don't think so,' he says. 'I heard it last night as well.'

We listen in silence as the aeroplane goes past.

'It's almost like an influx or a flushing of the system, or something,' Dave says. 'It's that vibration, that heavy rumble vibration.'

We sit for a while. I watch the Founder smoke a cigarette. Then I eat a chocolate Hob Nob and try out some experimental amateur origami with a corner of the local newspaper. Then I get up and look out of the window for a bit. Then I sit back down again.

'So, what do you usually do when you're here on your own?' I ask him.

'I don't just sit around doing nothing, you know,' he says. 'I'm a very accomplished guitar player.'

'I know,' I say. 'But what about the ghosts?'

David takes a puff of cigarette. 'I have a plethora of research data and information that I work through.'

'Do you do experiments?'

'Of course. Recently, I've done experiments where I pretend I'm slapping someone, like this.'

I watch the Founder stand up behind his chair and strike an invisible human across the face. He does it with his fag hanging off his lip and one eye closed against the smoke.

'So, when I'm a ghost and someone is actually sitting here, that's going to happen. And another thing I have experimented with is this cup. I lifted the cup like this and I moved it off the table and round in the air. So in two or three hundred years, that cup's going to do that and they're all going to turn around going, "Oh my God! Did you see that?" I do it with cushions and broom handles, too.'

I picture David by himself, in the night's cold, pin-prick hours, walking around the house, hitting imaginary future-people and making broom handles float. For some reason, it suddenly makes me feel very sad.

'Would you like to haunt this house when you're dead?' I say.

'I would,' he says. 'Yes, I dearly would.'

There's a long silence. It's after three in the morning now, and the house has been at peace with itself since the locals left.

'Do you believe in heaven and hell?' I ask.

'No,' he says, 'people were seeing ghosts BC – Before Christ – so that rules that one out totally. The earth is millions and millions of years old. You know, I think the Bible is a damn good book, but it's nothing like the original translation. How can we translate that when we still have difficulty translating the original Latin, which is only five hundred years old? It's very difficult, because it has so many syllabuses and nouns and whatever. It's like the voices on the EVPs I've recorded here. Most of them are in German Latin, which is what people spoke until the nineteenth century. It wasn't until eighteen-twenty-something that we began to speak English. A lot of people don't appreciate that fact.'

The Founder looks at my failed origami, open local newspaper, biscuits, notepad and pens. 'It would probably be advisable to put this stuff away,' he says.

As I tidy up, I decide to make idle small-talk about vampires. I wonder, as I clear the table, if the one that he told me about in Wales, last winter, was the scariest thing he's ever seen.

'Not scary,' he says. 'I wouldn't say scary because that's the one thing I don't do.'

'What was the most daunting thing, then?' I ask. 'The vampires? Or perhaps that glowing thing that came to you in the night?'

'Yes, it could be one of those,' he says. 'Or it could be the lycanthrope.'

'Lycanthrope?' I ask.

David sits down, opens his cigarette packet up again and pulls one out. 'A werewolf,' he says. 'It's a technical term. This woman just got down on all fours and started howling and hissing, and all this hair started growing out of her back. And it was really, really odd because her fingers became elongated, and then the nails came to a point and just grew.'

I sit up in my seat and tighten my mental seatbelts. 'Wow,' I say. 'That does sound really fucking daunting.'

'The confusing thing about this case was that she was supposed to be the mother, but it didn't all tie in. It was all confusion.'

'Sorry?' I say. 'I don't … '

'The mother said she was the mother and yet all the photographs of the mother didn't look like the mother.'

'Mother of who?'

'Of this couple.'

'So, you're saying that –'

'She just didn't look like her likeness in the photographs. You can tell that. It doesn't take a genius or Einstein to work that out. And the other odd thing was they had photographs of children but there were never any children in the house.'

'So this couple,' I say, 'these werewolves, they called you in to investigate a ghost case?'

'Yes, but it wasn't a ghost case. They just wanted to play with us. They were probably just bored of the usual … whatever it is they do.'

'So, what triggered her off then?' I say. 'Did you anger her?'

'No, I was very polite. I always am. This is one of those things I don't really like talking about too much. It brings back all these memories of what happened that night.'

I take this as my cue to stop asking. We sit quietly as I drain the dregs of my tea. Then, over the rim of my cup, I notice that the Founder is still looking at me, expectantly. I look back at him and raise my eyebrows.

'But, yeah … ' he says. 'In front of my eyes she started turning into a werewolf.'

'So, what were the couple doing while the mother was turning into a werewolf?' I say. 'Did they start changing, too?'

'Well, no. They were vampires and she was a werewolf.'

'Oh my God!' I say.

'I know,' says David, tutting. 'Talk about keep it in the family.'

'Did she attack you?'

'No, I calmed the situation very, very quickly. In situations like that, you have to do something very drastic.'

'What did you do?' I ask.

'I told a joke.'

I watch the Founder take a final draw on his cigarette, which he stubs out carefully in the ashtray. 'Do you remember what the joke was?'

'Well,' he says, 'while she was growing her hair and hissing and everything, she said, "I don't like you" in a gruff man's voice. And I turned around and said, "Don't worry, love, not many people do."'

There's a silence.

'And that joke reversed the werewolfing process?'

'Yes,' he says. 'All her hair shrunk back and the situation became normal.'

I look at the Founder and the Founder looks at me.

'Dave,' I say, 'I think I'm ready to go to bed now.'

As I brush my teeth, I start to wonder if the plan I made back at Michelham, to simply believe *everybody* that I was going to meet has, in the end, proved a little naive. But then, tempting as it is to let men with werewolf stories persuade you that there's no such thing as the supernatural, I think it's a mistake. Take this place, for example. I have no doubt that this house is haunted. There have been quite a few moments since my arrival that David and I have simultaneously jumped and looked around towards the same direction or doorway. And we have seen the same things – a black flicker, a movement in the air. The difference is, whereas I am unsettled and curious about these incidents, the Founder is rarely in any doubt that we have just been visited by an angry cavalier or a fat old monk.

David and I go to bed on two settees in a newer, unhaunted part of the house. That night, I find it very difficult to sleep. And not because I'm daunted. The Founder, you see, insists on sleeping with all of the lights and the television on.

'I'm not scared of the dark,' he tells me when I ask him why, the next day. 'By any means whatsoever. In fact, I don't really need the lights on at all. It's just nice to have something on in the background, like the TV, or something.'

After a mostly uneventful day, we go to bed again. And, once David's audibly asleep, I hop across the carpet in my sleeping bag and switch the TV and the lights off.

I lay my head down, close my eyes and, just as I'm slipping comfortably down the slope, I'm suddenly dragged back into full consciousness.

'You frightened the crap out of me,' says the Founder. He's out of bed and turning all the lights and the television back on. He's wearing tight black pants. He has a tattoo.

'We have to have them on for security purposes, I'm afraid,' he says. 'So potential intruders think that there are people in.'

'Can't we turn them on in another room?' I ask.

'No, it has to be this one,' he says, before getting back under his quilt on the sofa across from mine. Within seconds, he is snoring again.

The next day, we get to leave the house. The Founder has been telling me about the food in the local pub. He says that not only are the meals reasonably priced, they are almost unbelievably delicious. 'As good as you'd get at the Savoy,' he says. They serve coffee with a little jug of cream and a chocolate on the saucer. And the plates, he says, measuring out a length roughly the size of a manhole cover with his hands, are 'this big'.

Before we leave, David sets up his Minidisc recorder to catch any activity that might occur in our absence. He's meticulous. He makes me witness the unwrapping of a blank disc and watch as the machine's display flashes up 'Blank Disc'. I then have to record an intro, with a time, a date and an estimated time of return into the remote microphone.

It's just gone 6 p.m. when we reach the pub, which is warm, empty and looks as though it has been designed in a 'modern contemporary style' by a TV makeover crew on a tough schedule. As I'm trying to decide between chips and onion rings to go with my steak, the Founder receives a call on his mobile from his friend Paul, who used to be one of the paranormal experts on *Most Haunted*.

'Paul's a bit worried,' he tells me, once the conversation is over, 'about Living TV.'

'Why?' I ask, putting my menu down.

'Well,' he says. He looks around him, even though the pub is empty. 'You know how they've got all those gay shows on there? *The L Word* and *Queer Eye For the Straight Guy* – all that stuff?'

'Yes,' I say.

'Well, he's worried that they're going to do a … '

'A gay ghost show?' I say.

'Exactly,' says the Founder. He puts his menu down and looks around him again, conspiratorially. 'I mean, it's obvious really, isn't it?'

'Yes,' I say. 'They're bound to just bring the two concepts together.'

'Right,' he says.

The waitress comes over and we order our food. I read the menu again for a bit. Some time passes.

'I don't get paid for house-sitting, you know,' he tells me, suddenly. 'I don't get nothing, except for a litre-and-a-half bottle of Southern Comfort.'

I look at David and smile. My brain is empty again. I begin to feel a bit guilty about it.

'I don't even drink,' says the Founder. 'Well, hardly ever. But he always brings me the same thing. I've got a row of bottles of Southern Comfort in my bedroom at home. I'll wait until I've got a caseload and take them to an off licence and say, "Here you are. How much do I get for those?"'

'That's a good idea,' I say.

I pick up the laminated menu and examine it quizzically. I straighten my knife and fork in front of me. I arrange them exactly one thumb-width from the edge of my place mat, which is dead straight, and precisely a fingernail's distance from the edge of the table.

'Don't get me wrong, I do like Southern Comfort,' says David. 'But one bottle of it lasts me four years.'

And then, the food comes. It's served on enormous plates, and it is delicious. We eat it, happily and with slurps.

After our elaborate coffees have been drunk, we troop back

through the freezing night to the house. There's a full moon. I don't know if it's the monochrome lunar light that's glowing over the roof, walls, bare tree-branches and wide empty fields, or whether it's the age of the building or what. But as we walk up this remote country lane, the scene in front of us is truly eerie. And not in a cheesy ghost-train way either. The ancient little medieval home at the end of the path looks foreboding, brooding, unhappy. Its blank white walls and mean little windows give a void, pallid and deathly sense and some sunken, primeval instinct is telling me to turn around immediately and leave.

We walk into the house and the old wooden door, with its heavy iron bolts, is slammed and locked behind us. We take our coats off and trot upstairs to retrieve the Minidisc player. The recording doesn't cheer me up much. We listen to it sitting at the big table in the main room, with fresh teas. We get past my intro and the sound of us leaving the house and, almost straight away, the latches start going. And then there's footsteps and crashes and strange swishing sounds.

'The first time I heard this sort of thing,' he tells me, 'I was straight out of the door and down the pub. I couldn't come back in for three hours.'

'What is it?' I say.

I suddenly feel impotent and abandoned. What would Amorth say about these noises? That this is Satan's work? That flocks of demons are dancing through the dimensions and demonstrating that this house is in the darkness, well outside of God's kingdom? Or could they be Stone Tape sound recordings? Quantum bunches of soul? Rogue radio waves? Or a fault with the recorder? Or insects? Or … or … or … I sit there stilly and all these thoughts coalesce into a terrible theory-twister that ranges through my brain. My head drops into my hands.

And then I remember what I've done. I close my eyes and wish, *wish* I hadn't. Earlier on, I insisted that David let me sleep upstairs, tonight, in the 'haunted room'. I sit and sulk at myself quietly for a time.

Eventually, David starts to tell me about his life, and about his relationship with his fiancée that broke down fifteen years ago. He starts looking sad, so I try to cheer him up by mentioning the good things he's got going on. His ghost club, his new ITV1 show and, most of all, this place. He's so obviously passionate about it and truly happy when he's here. I say, 'It's almost as if your love affair is with this house now, isn't it?'

'It is, yes,' he says. 'I am in love with this house and I think if I was unable to come here again I'd be totally devastated, just like I was then. It would be the same feeling. A feeling of dread. That's a nasty feeling. Your stomach sinks, doesn't it?'

'It does,' I say. 'It's bad.'

I smile at him, encouragingly. I wonder if this is the moment to go further, to try and get to the bottom of his unusually intense feelings towards this house.

'I was thinking, Dave,' I say. 'I get the impression that … do you think that you lived here in a past life?' I say.

'It's quite possible,' he says. 'Yes.'

'I also get the feeling that maybe you think you used to own this house? Like, this house is actually yours?'

'Yes,' he says. 'That's very perceptive of you. I do have a sense that, to me, this is home. Someone did say that I looked like one of the faces that has been seen in the window over there.'

'You told me that you play the lottery religiously,' I say. 'Is that because you hope to buy this house? To reclaim what's yours?'

David picks up his cigarette packet. He puts it down. He stares into space between his head and his hands.

'It would be nice, yes,' he says. 'It would be very, very nice. I do think the person that is deserving of it is myself.'

'So, if you used to own this place, some of the ghosts in here could actually be you,' I say.

'Yes,' he says.

We sit in silence for a while longer and, eventually, I take my sleeping bag up the stairs and into the haunted room. It's horrible in here. It's as if the air is made of invisible sponge. As soon as I walk in, I feel forcibly bounced out again. David describes it well. He says it's 'a room that doesn't like people'. And it's freezing. The ceiling sags down with age, the beams are jagged and crotchety and the windows are warped and thin. Outside, the moon hangs above the ancient winter trees. I bend down to unzip my sleeping bag and the moment I do, become convinced that somebody is behind me. I turn. Nothing. I start to hum to myself for company, but can't rid myself of the feeling that I am being watched. It's as if the room is furious at the impertinence of my presence. There's an overwhelming feeling that something is terribly wrong. I hum some more, self-consciously, and close the door. And then I stand, rigidly, watching it. I'm convinced that the latch is going to open, just like it did before, with the old lady. I wait and breathe. Eventually, when I've sufficiently gee'd myself up, I turn off the light and climb, slowly and with as little noise as possible, into my sleeping bag.

The darkness curls around me. It seems, somehow, bigger and older than normal night. The room creaks and sits, as if satisfied now that it's in its favoured state – cold and black and pitiless, like a coffin. I lower my chin into the zip of the bag. Outside the window, the fields and the trees are silent. Inside, I can see floorboards and moonlight and ...

Something. I can hear something. Breathing. And it's not mine. My eyes stop moving. I'm scared and thrilled and rigid. In, out. It's there. I heard it. In, out. It is there. And it's not mine. I hold my breath and look and listen and try to hear it. It seems to be coming from the armchair, behind me. I can't move to look. I can't move at all. The floorboards are hard under me and my toes are cold and I'm suddenly very aware of the parts of me that are exposed to the cold air – my neck and my head and my nose – and I'm scared. I halt a whimper in the back of my throat and listen again and what's there, what is it? An apparition? A black shadow?

'Who's there?' I ask, weakly.

It stops.

And I bolt.

Out of the door and down the steps and round the corner, my sleeping bag trailing after me, my socks falling off my feet.

'Fuck!'

'I wondered how long you'd last up there,' David says, peering up from out of his quilt in the blazing light of the unhaunted living room.

'There was breathing!' I tell him. 'And it wasn't mine!'

'Was there?' he says. 'Great. I'm amazed you lasted as long as you did, actually. Most people are out of that room within seconds.'

The Founder puts his head back down again. I stand still for a moment and look at him. I feel anti-climactic. This isn't fair. This is my big moment. I try to prolong it.

'It's horrible in there,' I say.

David doesn't lift his head from his pillow. 'It is,' he says. 'I can't go in there for long.'

'Has anybody heard breathing before?' I ask.

'Yes,' he says. 'That's been heard before.'

'Often?'

'Not very often. It has been documented by visitors who've come here. There was one girl who heard heavy breathing followed by the words "get out of this room". She was quite upset.'

'It was incredible,' I say.

And it was. But as soon as I lie down on my sofa, my brain starts trying to rationalise it. The first query it slips me is: was the whole thing a dream? But I'm sure I wasn't asleep. Next, I remember that Ian Wilson, the author of one of the books I read at the British Library, heard ghostly breathing in a haunted bedroom. So did Jung and Lou Gentile. Coincidence? Or did I have some sort of auditory hallucination triggered by the memory of those things? What if Dr Mark was right? What if I heard the trees in the wind and fear pushed me into an intense, super-aroused state and my mind jumped to a wrong conclusion?

And then I start berating myself for not having the courage

to look at the source of the sound. What if there was an actual, visible ghost there? What if a fully formed black shadow, or a detailed apparition or a demon was standing there and terror made me miss it?

The next morning dawns pale and dank and ordinary. But as the Founder drives me to the station, the world, to me, seems more alive and magical and bewildering than it has for a long time. Now, more than ever, I'm sure that there is more to life than love, money and worrying. There is more to reality than we can see. And that's the best news that any human can have. I look out of my window. The car passes a bus stop and a supermarket and a dark stone church with a tall spire. It has a fluoro-orange poster with some pro-Jesus propaganda written on it. I follow it with my eyes, my head craning, until we pass it completely.

'Every time I leave that house, I get a tear in my eye and a lump in my throat,' David is saying.

'But when you die,' I say, 'you'll be there for ever. And then you can meet all the ghosts that haunt it and hang out with them.'

'No,' he says. 'I'd probably not be able to see them because there's a difference in time periods.'

'Oh,' I say. 'Won't that be quite lonely? Not being able to see anyone else for all eternity?'

The Founder looks at me, his eyes big and heartbreaking behind his fragile glasses. 'No,' he says in a small voice. 'I think that it would be quite nice.'

'I got one for you.'

'I'm sorry? Who is this?'

'What time is it over there?'

' ... um ... '

'About four-thirty a.m., right?'

'Yeah, I ... is this Lou?'

'Listen, Will. I got one for you. I'm going to need you to be at Philadelphia airport within the next twenty-four hours.'

'What? That's impossible! Twenty-four hours? I can't be –'

'We'll be going to Kentucky to see a boy who is possessed and we'll be going to his house where they've had black shadows, knocking, banging, voices, faces in mid-air. We're going to be driving all the way, all night, and this time, you gotta follow the rules. Make sure you stay with me and DO NOT provoke anything. I already have the demon's name and things could get real bad.'

'Lou, I'm sorry. There's no way I can just get on a plane and be in Philadelphia in twenty-four hours.'

21

'And that's God?'

Beyond the quiet rumble of the car engine, there's another, stranger sound. I first heard it when we last stopped, about 150 miles ago. I try to pick it up again now, through the steady hum of cruising tyres on interstate tarmac. The coffee dregs, inside two big-gulp beakers that are held from the dashboard in special cup-holders, are now cold. Their smooth, liquid surfaces pitch back and forth in time with the rhythm of the engine. The rosary that's hanging down from the rear-view mirror is also swinging to the beat of the road. Every now and then it hits the glass of the windscreen. It's hypnotic.

'There's a battle going on,' says the demonologist. 'There's a battle going on between good and evil. Between things that are in the light of God and things that are out of the light of God.'

I reach down between my legs, pick up another Hickory Smoke Jerky Stick, rip it open with my teeth and bite down.

'You're always going to have people who say, "There's no battle,"' Lou says. 'But I've experienced waaayyyyy too much stuff to even think about saying, "Oh, no, there's no battle." There's a battle.'

And then, the fog rolls in. Out of an empty sky, the twisting mist rushes quickly towards us, opens up and scoops us in. And all the while, the drum and rock of the car doesn't falter.

'Right now,' Lou says, leaning towards my tape recorder that's on the ledge behind the steering wheel, 'Will and I are driving through a big cloud.'

'Shit,' I say. 'Lou – look at the time.'

The red digits on the dashboard read 3.33 a.m. Lou takes it in and thinks for a beat.

'Hmmm. You know what?' he says. 'I ain't taking any chances either.'

Lou lifts his petrol-leg a little. The engine drops a pitch as the car slows down. He looks at me sideways and quickly from behind lowered eyebrows.

'We haven't encountered anything strange, except this,' he says. 'We're driving through a big cloud, even though we just went through about sixty mountains and didn't see anything like this.'

Now that the car's incessant drone has quietened, I push my ear into the window and try to hear the sound again. I think I can hear … something.

'That shit you were telling me about Janet,' Lou says. 'That's some really wild stuff, man.'

'Yeah,' I say. 'Honestly. Her voice sounded exactly the same as your EVP.'

'And that one thing you said about when they duct-taped her mouth shut … '

'And filled it with water … '

'How about when she levitated in the room? She was fucking levitating! And I can guarantee you that shit still goes on, but she's not gonna talk about it because she doesn't want it to come back.'

'Do you know who she reminded me of?' I say, peeling back the clear plastic wrapper and taking another bite of the jerky stick.

'Kathy!' Lou says.

I look at Lou, and nod. He laughs triumphantly and hits the steering wheel with the heel of his hand. The clock flicks to 3.34.

'You see?' he laughs. 'I'm telling you, man! You must have been going nuts!'

'Honestly, Lou,' I say. 'She gave me this look. It was just the same as the look Kathy gave me.'

Lou laughs again. A big, rolling guffaw that fills the car and ends with a 'woah!'

'And she says that there was something still there,' I say, 'until her mother died.'

'It's there,' he says, nodding. 'Oh, it's there all right. That case is too fucking powerful.'

'She also told me that she did a Ouija board with her sister, before it all kicked off. And when I told Amorth, he said it was the devil.'

Lou stares straight ahead, at the night and the miles in front of us. 'Oh yeah, they pulled the devil through,' he says. 'Not *the* devil. A devil. Anytime you use divination like that, forget it. You're going to wind up with something bad. Ninety-five per cent of the cases I get involved with are to do with Ouija board involvement. And I'm not talking about "ghosty" ghosts, like apparitions. These are *bad* cases. But you gotta understand, I'm not saying that everything is demonic. You have to understand that. You don't just walk into somebody's house and say, "This is a demon." It really takes a lot of research. I'll give you an example. It's like this case we're going on down to tonight. This one's different. There's just too many signs with this one.'

Lou hunches down and moves his head forward to peer through the fog.

'Man,' he says, 'I can't see where the hell we're going.'

The rosary hits the windscreen, again.

'Do you think this fog is anything, y'know, bad? It came at exactly 3.33.'

'Well, we're now in Kentucky and all this way we didn't see anything like this. But you got to look at the possibility that it's fog, and that's all. Nothing bad is happening and that's great. I'd rather have it that way.'

'You know when the Carvens' clock stopped at three o'clock?' I say. 'Do you think that was a sign?'

'Well, three *is* a sign of the demonic,' he says. 'But if that's the only thing that showed any sign of something demonic, there ain't nothing there. It's got to be a coincidence. It just has to be. At the beginning of a haunting, if you get three knocks or something three starts to show up and continues to show up, then that's a different story. But with the Carvens there was nothing really there that showed a sign of being demonic.'

'Did they win their court case?' I say.

'I don't know,' Lou says. 'I kinda stay out of the legal thing because, you see, when you get involved with a case and it becomes legal, people want you down there at a specific time and if you can't make it, you're an asshole. But they were happy with my analysis. I didn't try to blow smoke up their ass. I'm not a big proponent of orbs and globules. But when you see, which you saw, that globule follow her up her stairs, what can you do?'

'I emailed a parapsychologist about that,' I say, 'and he said it was just insects.'

'There were no bugs in that house,' says Lou. There's a silence. 'Well, there could've been. But what are the odds that a bug, out of focus, is going to follow Mrs Carven up the stairs, and then, when she turns, you see it go towards the haunted room? It wasn't insects. It wasn't pollen. It wasn't dust. There was something in that house.'

'And in Kathy's house there were so many of them,' I say. 'And I looked in the kitchen on the night. There wasn't a swarm of insects there.'

Outside the car, the Kentucky night is clearing. Inside, I'm becoming restless. The gallons of truck-stop black coffee that I've drunk to saturate my jet-lag have drained their way downwards. The feeling is heavy and insistent and uncomfortable.

'Could we stop for a minute?' I ask Lou. 'I have to piss.'

'Sure,' he says.

He indicates in the empty road and pulls onto a verge. As soon as the engine stops, I hear the sound again. It's incredible. But, stepping out into it, it's also unnerving. You see, you can't find its source. All you can tell is that it's coming from a multitude of beings. Their presence is invisible, and yet it commands the landscape. The people who live out here, the folk of the sparse settlements of Bible-belt America, hear it all the time. But to me, as I stand by the cooling, ticking car, pissing into the long grass, it's startling. I have to remind myself that I have nothing to be scared of. After all, it's just insects. Crickets, to be precise. Swarms of swarms of them. Nothing weird or unexplained. In this part of America, their chirruping, two-tone hymn fills the air every night. It's a suitable soundtrack for the stars, the panoply of galaxies that glimmer above my head. I look up. Now up there, there *is* something weird and unexplained. The bright, distant universes are incredible, unbelievable. But, as unbelievable as they are, they're a fact. And only because they're plainly visible up through the night sky, our widescreen windscreen out into the universe. If you couldn't see it, you wouldn't believe it for a minute. And if I couldn't hear the crickets, I'd never think I was surrounded.

'You ought to be careful,' Lou says as I get back in, positioning my feet around the rubble of snacks and Pepsi cans that I've dropped around my feet.

'That place you heard the breathing in?' he says. 'That place is obviously totally demonically infestated.'

'Well, I'm OK,' I say, as he turns the ignition and curls his fingers over the gear stick. 'Amorth gave me a once-over. I'm not possessed. Nothing followed me home. And ghosts can't kill you, can they?'

'Well,' he says, 'if objects can be thrown about or levitated, what's stopping a knife, or an object that's pointy, from flying out and hitting you in the jugular vein, causing catastrophic injuries resulting in death?'

'I hadn't thought of that,' I say.

I look out of my window for a while and my thoughts drift. We roll on in silence for ten miles, twenty ...

'I was thinking about that demon you saw,' I say, eventually, 'and I met this other guy, David Vee ... '

'Who?' says Lou.

'Um ... ' I reflect for a second. 'No, actually – don't worry ... '

We turn a sharp corner around a dark hillside and, as we clear it, the road ahead shows itself in all its length and grandeur. It stretches out languorously and turns and climbs for miles, all lit up and empty. We are now ten hours into our drive, to an outbreak of the 'battle' that both the demonologist and the exorcist talk about that has erupted in a family home right in the heart of holy America.

'When you're dealing with demons and devils,' Lou says, as we reach cruise speed, 'and you have somebody that asks you, "Hey, what does this name mean?" a lot of times, the name makes absolutely no sense. But when you have a situation that arises, like the one we're on our way to, where you have a kid, and his mother asks him, "Who are you?" and the kid utters the name Ogalegal ... well, it's very rare.'

'Ogalegal?' I say. 'Is that the name of the demon?'

'Well, Og is the name of an Israelite devil that's connected with breaking faith. But I was stumped when I heard "alegal". I thought, OK, a kid saying Ogalegal doesn't know what he's talking about. Then I thought, let's do this another way. What's this kid's birthday? It's the eighteenth of October. Now, every day has a specific demonic energy, at least, according to the old texts. So I looked up the eighteenth of October. What do you think I found?'

'I don't know,' I say.

'Egibiel. Og is the devil. Egibiel is the demon. Now, I saw this, and I thought, OK, say this name real fast. Ogebeable. Same thing. So don't be surprised if some very strange things wind up happening over the course of the next seventy-two hours. Because this right here,' Lou says, hitting the top of his steering wheel with his outstretched finger, 'is very, very rare.'

'How rare, exactly?' I ask.

Lou takes his hand off the wheel, for a second, to rub his rose. 'Man, it's rare to even find something close to a name that somebody repeats. A lot of people call me up and go, "Oh, yeah, I got Leviathan." Yeah, OK. Whatever. Yeah, sure you got Leviathan in your basement. Is it possible? There's a one in sixty-million chance that that would happen. In my lifetime, I may never encounter a true demon in the upper hierarchy.'

'So, this kid is definitely possessed?' I say.

'There's a lot of research gone into this. There's a lot of pieces of the puzzle fitting together. I've covered all the bases,' he says. 'I've spoken to the pastor. I have a psychological evaluation of this kid that's already been performed. I have a meeting, tomorrow morning, with the actual psychologist of this kid, and the mother. And we're going to be seeing him for two hours, at the institution.'

'Institution?'

'Yeah, the kid had to be locked up for his own safety. Now, there's got to be a reason for that happening.'

'Do you mean that there's a paranormal reason behind the kid being locked up?'

'Yes,' he says. 'Yes. There just has to be a reason for it. You see, when evil manifests itself, there's always something good trying to sneak its way in to the situation. Good always tries to overwhelm it.'

'And that's God?' I say.

'Exactly,' he says. 'It's divine intervention. You know, God is not an old man with a beard hanging out on the Planet Neptune. God is a positive energy source. An extremely powerful energy source – a human being cannot be in the presence of God without being destroyed. And sometimes, God intervenes. I've been on cases when things have been just about ready to go bad and this encompassing light that you can't describe – I call it white, but it's not white – it comes into a situation and it releases everything.'

'So the fact that the kid's been locked up,' I say. 'That's God? That's divine intervention?'

'Exactly.'

'And there's definitely nothing else wrong with the kid?' I say. 'Nothing medical, psychological?'

'No,' says Lou. There's a silence. 'Apart from the autism, nothing at all.'

'He's autistic?' I say. 'But ... Lou!'

'Will,' Lou says and looks at me, 'listen. If what I've been told is correct, even the kid's psychologist says this case is paranormal.'

As I'm absorbing this, I remember Father Amorth telling me that psychologists and doctors sometimes refer patients to him – and that he's cured them. Like the epileptic boy who, after his exorcism, never had another fit. Just then, we pull off the road, into another petrol station. It's almost four-thirty in the morning. I've been awake for a day and a half.

'Will,' says Lou, 'can you pay for the gas? Go up to the counter and ask for thirty bucks on pump five.'

I step out of the car onto the warm tarmac. The crickets are reaching the dramatic climax of their nocturnal opera. Up above me, traffic lights hang from wires. There's an articulated lorry parked at the side of the complex with curtains pulled over its windscreen. Blue and black flashes of a TV glow up onto the other side of the thin material, evidence of a world within a world that I don't want to know about.

I walk up to the counter.

'Can I have thirty bucks on pump five, please?' I say, to the woman sitting behind it.

'Wha'?'

'Can I have thirty bucks on pump five, please?'

'Eeerrr.' She pauses and looks at me. 'Wha'?'

'Thirty bucks,' I say, 'on pump five.'

'Yaw goin' a haff ta repeat that reeeaaal slow,' she says.

Lou walks up behind me. 'Thirty bucks on pump five,' he says.

Behind us, there's a man speaking loudly into his mobile phone. From the content of his conversation, it's become apparent that's he talking about a woman who has recently sold him some sex. He's describing her anatomy in a way that suggests that the person on the other end of the phone is also a customer. And a happy one, at that. The man on the phone is wearing nothing but a pair of jeans. They are slung half an inch below the level of his hairy arse-crack. He is bald and enormous and appears as if

God has shaped him roughly out of a huge lump of clay, with two careless squeezes. In the passenger seat of his car sits his girlfriend. She's drunk and hawwing sloppily into a mobile. As the man approaches the counter, it becomes clear that he is also drunk. He walks like a marionette. At this moment, I am so scared of this man it feels as though I'm in danger of liquidising on the spot and trickling into a nearby gutter.

As he fumbles with his dollars, I slip into the toilets around the back. I stand at the urinal and exhale and watch the entrance. After I hear his door slam and the car drive off, my bladder allows itself to release, and my attention is taken by a sticker on a condom machine. It contains the official small print, amongst which is this caveat: 'Condoms do not offer 100% protection from HIV/Aids. The only guaranteed protection is a loving, heterosexual relationship.' Somebody has underlined the word 'heterosexual' so deeply that the plastic sticker has split open, and the paint underneath it is scratched to the metal.

'We're dealing with Baptists all weekend,' Lou tells me as we drive off.

'Lou,' I say, 'you know you said that –'

'You're still worrying about the autism thing, aren't you?' Lou says. 'Well, listen. Before moving into this house, Denzel was a completely different person to who he is today. He was still autistic. But the way these kids are, they're predictable. Their attitude, their constitution, their whole mental psyche is usually one way. You can expect how they're going to react to certain things. It's always the same. But when they moved into this house, Denzel started to say that he would see somebody coming out of the closet. He would always want to sleep with his mother. She would go, OK, he's just a kid, it's no problem. But after a while, it became old. So, finally, the mother slept in there with him, and she saw what appeared to be a black shadow. Well, she ran out of there and started screaming the Bible.

'After this, Denzel started to become very violent. He started to draw pictures of what looked to be some kind of monster.

Then, the mother would hear as though somebody was literally walking on top of the rafters in the middle of the night. All the while, Denzel's mother was talking to the psychologist and he couldn't figure out why this kid started displaying signs of spiritual-slash-physical torment. So she went to her pastor and he did an exorcism over him and Denzel starts freaking out. He started spitting and cursing and saying a lot of things that were not befitting a kid of his age. He came out with information that he didn't know and that's a sign, right there, that we're dealing with something demonic. And, anyway, this local paranormal investigator that was called in by the mother – he was only there for three hours, and he saw black shadows, he heard things and saw things moving around. That's when he called me in. And it's his house we're going to first.'

As the pale-blue dawn begins to glow down the telegraph poles and the road signs, huge rectangular buildings emerge from the black shadow of night. There's a drive-thru McDonald's with a huge neon 'M', a drive-thru pawn shop with a huge neon dollar sign and a drive-thru 'First Baptist Church' with a huge neon crucifix.

'What's the investigator's name?' I ask.

'Bob. But before we get to his place, I gotta warn you. My investigations into this situation have given me reason to believe that it's actually the mother that's possessed. She could be drawing attention to the kid so the spirit isn't detected inside her, if that makes sense.'

'So, the kid's the patsy?' I say.

'Yeah,' says Lou. 'Exactly. And there's one other thing. You can't tell these people what it is you're doing. You have to tell them you're my research assistant who's present only to document, you know, stuff for legal purposes. If I tell them that you're doing your own thing, it complicates matters. It gets people nervous. So, do not give them an inkling that you're from the UK.'

'But, Lou – what about my accent?'

'I know it's going to be hard, but ... here we are.'

The demonologist pulls up next to a black pick-up that's parked outside a modest bungalow on the end of a small-town street. The lovingly waxed vehicle has two bumper stickers. One has the words 'Marriage = 🚹+🚺'. The other says, simply, 'Proud American'.

Bob is thin, pale and buzz-cutted. He greets us in his dressing grown, the sticky fug of deep sleep still clamming up his face and slugging his movements. Bob yawns, apologises and shows us into his living room. He promises us that his wife will be up to make us 'a real good country home-cooked breakfast' once he's shown us his home movie of an autistic seven-year-old having a violent exorcism. It's the events of this film that convinced the people around Denzel that he was demonically possessed and led Bob to call Lou in.

'The local pastor that's been praying over Denzel – he's a real good guy,' he tells Lou after we've introduced ourselves and sat down heavily on his sofa.

Bob leaves the room for a moment, then returns, wheeling with him a small portable TV that's attached by wires to a DV camera. He parks it in front of us and bends over, the belt of his towelling dressing gown touching the carpet. Then, he presses the fiddly play button. The static blinks off and the screen shows us a pretty late-thirties African-American woman sitting on a flowery sofa in a tidy lounge.

'Is there anything you'd like to add?' a male voice says to her.

The lady smiles, joyfully and with teeth. She says, 'God is awesome.'

Bob nods his head in reverent agreement, opens a tin of chewing tobacco, takes a pinch and pushes it down between his left gum and cheek. The room around him is sparsely furnished. There's nothing much to see, except a bookcase full of videos, a leather three-piece suite and a picture of Jesus and Mary in a frame on the mantelpiece. It's painted in soft, pastel colours and their eyes are big and round and dewy and Disney. I try to relax, to loosen the sinewy apprehension out of my shoulders.

The pastor is a great, grim silverback of a man. His shoulders sag with the weight of his arms and the fly part of his trousers is stretched tight over his stomach's swell. He looks like a bigger, more monstrous incarnation of the drunken mountain man at the garage earlier on. We watch him on the screen as he sits on a stool in front of Denzel. Spills of excess backside gloop over the small circular seat. With a black leather-bound Bible open on his pudgy palm, he reads out favourite passages, before asking the boy, 'What does that mean to you, Denzel?'

I'm captivated by the child. He has a bad squint and a sad countenance. His big dark eyes look up at the pastor. He doesn't speak. He just looks away, towards his mother.

'What does that mean to you, Denzel?' the pastor says again.

The child shrugs his shoulders and puts a finger in his mouth.

'Does it mean that Jesus Christ is your Lord and Saviour?'

Denzel looks at his mother again.

Lou's watching all this with a studious frown, his legs crossed expansively, his hand clutching his chin. He says to me, 'Notice how he keeps looking to the mother for permission to speak?'

'Denzel?' says the pastor. 'Do you recognise Jesus Christ as your one Lord and Saviour?'

'Jesus?' he says. He looks mystified and slightly afraid. He doesn't know what he's supposed to be saying.

'He's reacting too quick,' Bob says.

I don't understand. I look at Bob, then back at the television.

'Denzel? Listen to me,' the pastor says. The holy man has a crop of black stubble for a haircut. His fringe is a straight line that runs low above his piggy eyes. 'Do you recognise Jesus Christ as your one Lord and Saviour?'

'Jesus?' Denzel says again, quieter this time.

'I'm going to say some prayers of deliverance,' the pastor tells Denzel's mother.

'I'm real curious to know what you're going to make of this, Lou,' says Bob, in the room.

The pastor puts his slab, white hand on Denzel's dark

forehead. His thick raw-sausage fingers curl weightily over the boy's small skull.

'In the name of Jesus!' he shouts. 'I bind you!'

Denzel's eyes get a little bigger. His hands begin to clench.

'I bind you!' the pastor cries again.

Just on the edge of shot, Denzel's mother picks up her Bible and starts reading a passage out loud.

'In the name of Jesus!' the pastor repeats. 'I bind you!'

Silently, the seven-year-old starts rocking at his waist, back and forth.

'I bind you!' shouts the holy man, louder now, over the rising chanting of Denzel's mother.

Denzel's rocking gets more pronounced. He starts to make a sound. 'De de de de de de de de de.'

In the living room with Lou, the blue light of early dawn has turned to an overcast white. Bob sucks at the tobacco in his jaw, spits an oyster of brown juice out into an empty glass and says, 'What are you making of this, Lou?'

The demonologist shushes Bob as, on the screen, the boy struggles to get away from the pastor. With a sudden jerk, Denzel falls to the floor. The pastor goes with him, his heavy round knees banging on the ground. The camera shakes.

'Grab his wrists!' he shouts as Bob comes into shot and pulls the boy's arms back. The pastor has his hefty, wobbling arms gripped around Denzel's squirming legs. His mother is singing a hymn.

'Come out!' shouts the pastor. 'Come out in Jesus's name!'

The mother's lungs open up in full gospel fashion as she sings, 'Jesus loves me, this I know, because the Bible tells me so.'

It's as if a storm has opened up in the middle of the room. In amongst the frilly upholstery, leafy plants and paintings of idyllic family scenes, there's a chaos of shouting, fighting and praising the Lord. Down in the centre of this violent bedlam, the little boy is pinned to the floor by two heavy men. And he doesn't make a sound. You can see his white trainers and his sticky-out ears and his blue denim jacket with a patch on the pocket. You can see him struggling on the clean carpet of his home, wordlessly.

'Come out!' the pastor shouts again. 'Come out in Jesus's name!'

The mother interrupts her singing, briefly, to cry, 'Jesus!'

And then Denzel speaks. 'You are a bad man,' he says.

In the room, Bob's eyebrows raise up. Lou's eyebrows scrunch down.

'I love you!' replies the pastor. He's worked his way up to Denzel's waist now. The back of his head is fat. 'Jesus loves you!' With his free hand, he's opened a small pot of blessed oil and is trying to put some on the child's head.

'Mum! Mum!' shouts Denzel. 'Help me, Mum!'

But Mum is busy, singing to Jesus.

'You are not a doctor!' shouts the boy. 'You are not a doctor!'

'I love you!' the pastor shouts again. His stonewashed jeans are being pulled down in the struggle and his shirt is being pulled up.

'I love you,' he shouts. 'Jesus loves you!'

'Mum! Mum! Stop it! Stop!'

'Jesus!' shouts Mum.

'I don't want Jesus,' Denzel cries.

'You told me you loved Jesus!' the pastor cries. There's a menacing triumphalism in his voice and there are spots of spit on his lips. And there's fear in there as well.

'Shut up!' Denzel shouts. He's angry now. 'Shut up!'

In the back of shot, a man in a dark T-shirt is making coffee in the kitchenette. He appears entirely unconcerned at the bubble of hell that's popped open on the living-room floor, just a few feet away from him.

'The pastor's a real nice guy,' Bob tells us. 'He knows his Bible real well.'

'Motherfucker! Motherfucker!' Denzel shouts.

'Demon of profanity leave this child!' the pastor shouts back.

'Jesus died on the cross for Denzel,' his mother announces, in between caterwauled verses.

'Mother ... fucker,' says Denzel. His voice has changed. It's crumpled and teary now. Defeated. 'I don't like you,' he says.

'Demon of attention deficit disorder, leave this child!' says the pastor.

'Shut up, motherfucker! You are a bad man!' Denzel has got his shout back a little. He hisses as he struggles.

'Hissing is extremely common in cases like this,' Lou says.

Then, the boy spits.

'You see?' says Bob. 'There's a sign right there. And the profanity ... ' He spits another glob of slippery brown tobaccy-juice into his glass.

'Demon of I don't love you,' says the pastor, lying on the floor on top of the boy, 'leave this child!'

'I'm not playing with you, motherfucker,' Denzel shouts from inside the knot of limbs that are holding him down.

'Demon of retardation, leave this child! Demon of schizophrenia, leave this child! Demon of selfishness, leave this child! Demon of madness, leave this child! Demon of autism, leave this child!'

'Motherfucker!'

'Demon of profanity, in the name of Jesus Lord and Saviour, leave this child!'

'Mum! Mum!'

'Jesus loves me this I know! Because the Bible tells me so!'

'Mum! Stop it!'

'Watch his eyes,' Bob says. 'They turn black.'

'Lou,' I say, quietly. He looks at me, and I motion, with a small head-nod, towards the door that leads to the garden.

'Do you mind if me and Will step outside and have a smoke?' Lou says.

'Sure,' he says. 'Be my guest.' He gets up and pauses the video in his dressing gown and white socks. 'Anything you want, you just holler, you hear?'

There's a menacing quality to Bob's razor etiquette. It makes me feel profoundly unsafe. Perhaps, I think, it's just his down-home *Deliverance* twang. But, no – I think my discomfort comes from the instinctive feeling that the man who expends the most effort on the presentation of his personality is the man who's got the most to hide. And when the manners are this pristine, you've got to wonder about the private realpolitik that they're shielding. I smile at Bob as I step out of his French windows but I fail to manage eye contact.

Outside the house the crickets have quietened. It's almost as if they've been watching us, listening in on the guilty scene. They've been hushed by the horror.

'Well, I'm not seeing any signs of supernatural strength there,' says Lou. 'Or any facial changes. The change in his personality was really something, though. And there was spitting. And hissing ... '

'Lou!' I say.

'What?' he says. 'What did you make of that?'

'Where I come from,' I say, 'that's child abuse. Bob and that fucking pastor would be locked up if Social Services got hold of that video. It was horrific. That poor kid.'

There's a silence as I look at Lou and watch this sink slowly in. 'Child abuse, huh?' he says.

'That kid was petrified. That's why he was swearing! He didn't know what the fuck was going on.'

'These are all valid points you're making,' Lou says.

'He's not possessed,' I say. 'He's autistic.' There's a pause.

'All right, listen to me,' he says. 'You might be right. There may be nothing wrong with the kid. It might all be coming from the mother.'

'What? You still think this case is paranormal?'

'Sure,' he says. 'Even the kid's psychologist thinks it is. And he knew the name, Ogalegal. I'm telling you, man, something's going on. And that house? It's definitely haunted. When we go back in there, let me do the talking. All right?'

'Right,' I say. 'Fine.'

Back inside, we're presented with confusing breakfasts. There are sausages that are shaped like hamburgers, scrambled eggs with sugar in them and 'biscuits' that are actually scones.

'Shall I put the tape back on?' Bob says, settling down with a large oval plate on his lap.

'How much longer is there?'

'About, I guess, another three hours.'

I almost choke on my sweet eggs.

'Actually, do you know what?' Lou says. 'I don't want to see any more.'

'So what do you make of it, Lou? Pretty nuts, huh?'

'Well, I'd say that this is borderline spiritual warfare ... '

I clear my throat, noisily and obviously.

' ... borderline child abuse,' he says.

'Child abuse?' Bob says, his fork in mid-air, his thin lips slack.

'Be very careful who you show this tape to,' Lou says. 'As a friend, as a fellow ghosthunter, I'm telling you this should not have happened.'

'Oh, Lou, don't tell me that,' says Bob.

'I'm telling you, man,' Lou says. 'A judge would say this tape is child abuse. If I were you, I'd tell Denzel's mom that this tape got lost. I wanna see that motherfucker burn. I want it melted.'

At just gone 10 a.m. that morning, there's a videotape barbecue on Bob's patio. And when we finally arrive at Denzel's house, his mom has a hard time believing Bob's lie.

'My kid just chewed it up,' says Bob, shrugging limply.

'How is that possible?' she says.

'Well, I don't know,' says Bob. 'He destroyed the camera, too.'

'Isn't that a little bit strange?'

'Mrs Clarke,' Lou interrupts, 'I have a whole battery of questions that I need to ask you and I want to get through them as fast as possible. Can we sit down?'

We all move to the room where Denzel was sat upon. The day after the exorcism, Denzel had what Bob described as a 'very bad episode' at school. He ran out of his classroom and into the road. This was why he was sent to the institution. Of course, Denzel's guardians think that it was the devil in him that made him act this way. It hasn't occurred to them that the wailing chaos, the frightening and confusing and brutally weird scenes that happened around him and to him, for more than three hours, might have cracked his already brittle mental state.

During Lou's interview, Denzel's mother tells us about the things that are in her house. There are rappings on the wall. There are footsteps. There are disembodied voices. There are white balls of light and streaks. There are times when 'it gets really cold; it feels like a breeze is going by'. She sometimes feels prodded. And there are black shadow apparitions. She tells Lou that she believes in the devil. She says she 'knows he exists because it's a known fact that he exists'. She says that 'faith is believing'. Then she admits that some time ago, she used a Ouija board with some friends. The planchette moved by itself, with nobody touching it.

If Father Amorth were here, I think, as she speaks, he'd probably agree with Lou's diagnosis. It's the mother who's infested. The kid's just the patsy.

Halfway through the interview, Denzel's psychologist walks in. A middle-aged man with a tucked-in shirt and shiny leather shoes, he listens to the proceedings uncomfortably, his foot twitching irritably and his hands gripping the arms of his chair.

When it's his turn to speak, he tells Lou, 'Denzel has been diagnosed with a disorder called Aspergers, which is similar to autism. Now, I have a lot of experience in this area. I started off working with kids with autism and autism-type disorders in nineteen seventy-five,' he says, 'and Denzel's behaviour is consistent with the behaviour that these kids show.'

'Let's say,' Lou says, 'a situation was to arise where he was confronted, and constantly hammered with questions. Because of the autism, would he be on frequency overload and would he just, um, spaz out?'

'He would have problems processing that and coping with it.

He would become over-stimulated and the more over-stimulated he is, the more disinhibited he is, the more chaos he causes around him until the behaviour increasingly breaks down. If there is any kind of paranormal manifestation here, I would consider that as a stressor that would cause him to be disinhibited and act out. Denzel says these things are happening to him and I accept that. But I have not seen, from my conversations with him, anything that I could discern as a sign of any kind of evil in him. I just don't see that in him. When Mrs Clarke talked to me about this before, my impression was that it was more connected to the house and not coming from him. But I have no experience in this area. I don't know what is possible and what is not possible.'

Lou looks at him and frowns. It appears that the demonologist has been misinformed about the psychologist's opinions of Denzel.

'So in other words,' Lou says, 'you've not seen any signs of supernatural powers in Denzel? You've not seen his eyes change colour?'

I watch Lou earnestly quiz the psychologist and silently will him to stop. Suddenly, it's excruciating.

'You've not seen him display forbidden knowledge?' he says. 'Or talk in foreign languages?'

The psychologist looks at him square in the face. 'I've not seen any of that in him,' he says.

Despite this, Lou remains determined to rule Denzel out of his enquiries. So we climb back into his car for the two-hour journey to the institution.

When we get there, we find the child's mother sitting in a small empty social room, on a circle of chairs that surround a low nest of grey tables. Denzel is sitting at these tables, his legs curled on the floor underneath him. He's drawing on a pad with a red felt tip. Mrs Clarke is determined to show us that her son is possessed, that he's in such a state of manic demonic preoccupation that he speaks evil words and draws horned devils obsessively.

'What's that you're drawing?' says Lou.

'A church,' Denzel says, his head tucked into his drawing arm, his eyes solid with the concentration.

'That's a very detailed church.'

'Well, thanks,' he says.

We watch Denzel draw a fat crucifix.

'What's the door like in your bedroom?' says Lou.

'It's a creepy door,' he says. Then, he stops drawing and asks us all, in turn, where we went to sleep last night. When he gets to me, I reply, 'In Lou's car.'

He looks at me, sadly. 'I went to sleep here,' he says, before tearing the top sheet off his pad and starting a new picture. Then he looks at me again. Something has puzzled him.

'Where do you live?' he asks. And then I realise. It's my accent.

'Philadelphia,' I say.

'Philadelphia?' asks Denzel, his mouth gaping with disbelief.

There's a silence.

'Yes,' I say.

Thankfully, Denzel goes back down and continues with his drawing. Outside the door, two nurses stride past, their voices fading down the corridor. Through the small reinforced window, you can just see Bob, leaning against the wall, trying to peer in.

Then, Denzel's mother lurches towards her son and says, 'Look at that! What are you drawing?'

'Scary face,' he says.

'Why did you draw that?' she asks. She's excited now. Here comes the proof.

'Scaaarrryyyy faaaacccceee,' Denzel says.

'Where did you see that?' she asks, darting Lou a look.

'*Scooby Doo*,' says her son.

She slumps back in her chair, temporarily defeated.

'Do you get scared sometimes, like Shaggy?' Lou asks.

Denzel shakes his head.

'You're Mommy's brave boy, aren't you?' says his mum.

Denzel ignores her for a beat. Then he asks her, 'What makes you brave?'

'Jesus makes me brave,' she says.

'Jesus? Woaaahh! Jesus does?' he says.

'Oh!' she says, firing off another look in Lou's direction. 'Doesn't Jesus make you brave, too?'

'Yep,' he says.

Lou moves to get up. 'It was nice meeting you, Denzel,' he says, rising from his seat.

'It was?' Denzel asks, his big eyes peering up from his drawing.

As we walk out of the social room, I see Bob leaning next to a sign by a set of grey swing doors. It says 'Patients From This Unit Will Attempt To Elope Down Stairwell. Please Slam Door'.

'I can't take too much more of that,' the demonologist says. Something in his voice makes me look at him. Lou is close to tears. 'I feel like a fucking idiot,' he says. 'That kid's not possessed.' He pauses and rubs his eyes with the heels of his hands. 'But where did that name come from?'

'I don't know,' I say. 'Maybe it's a coincidence.'

There's a tense silence, as the three of us pace down the strip-lit corridor.

'No, there's something going on here,' Lou says. 'I just haven't figured it out yet. I don't think she's making stuff up about the house. I know there's something in that house. I fucking know it.'

'What the heck happened in there?' says Bob, following behind us as we go.

Back at Bob's, we sleep. And when the sun goes down and the crickets strike up again, we prepare to return to Denzel's house. Lou remains convinced that there's something demonic prowling around in there. Before he left, he tells me, he put a pinch of blessed salt in front of the closet door that Denzel and his mother have seen a black shadow coming out of. Holy salt, I remember, is Father Amorth's favoured method of dealing with a 'possessed house'. If there actually is anything demonically up with Denzel's mum, the theory goes that she'll have a rabid aversion to the stuff. It'll be like garlic to a vampire.

Half an hour later, as Bob and Lou and I turn back onto the freeway, I find that the demonologist is getting slightly more into his conceit that I'm his 'research assistant' than I am entirely comfortable with. Lou's leaning back in his driving seat and shouting to Bob in the back, 'Skippy here will be spending time in the room tonight. By himself. With the door shut. Right, Skippy?' he laughs as he drives.

'Right,' I say.

'Will's got the kid's room. As a matter of fact, I might lock his ass in the closet.'

'Ha, ha,' I say.

'Ah, he knows I'm just bustin' his ass,' Lou shouts to Bob. 'Dontcha, Skippy? But, seriously, you gotta remember my rules this time, Will,' he says. 'Believe me, you're not going to see Casper the friendly fucking ghost walking down the hallway. Something is going to happen to somebody. And when that happens, you must not freak out. Try to compose yourself as best you can. There may be knockings and bangings. There may be voices. Somebody may well get attacked. I gotta warn you, I don't think this is a ghosty ghost thing. This is definitely something along the lines of the demonic.'

When we arrive at Mrs Clarke's for our vigil, Lou gives Bob and me the same talk that I heard in the diner on the outskirts of Philadelphia. He runs through his rules and leads us, above the sound of the chanting crickets, in a rousing 'Our Father', which is supposed to help protect us from the oncoming demonic onslaught. On the long drive from Philadelphia to Kentucky, Lou told me that there was a battle going on, between good and evil. But we've found no evidence of the devil in Denzel. In fact, I think that if I have witnessed any evil today, it's in the very people who are convinced of their goodness.

But is there *really* such a thing as an independent force of evil? One that disappears up the ears of innocent sleepers, and causes them to wake up in the morning red-eyed and crazy, suddenly a thief or a wife-beater or a drunk or a paedophile? Is there, as the Christians suggest, something tangible floating about the planet, turning once-fine humans rotten? Could it be true that humanity doesn't own the dreadful things that it does? That wickedness is something that happens, not by us, but *to* us? Suddenly and without cause or context? I don't think so. Our behavior is defined by our personal history, our psychology and, perhaps, our genes – and the way those things pan out is down to the chaos theory of birth and life.

I think, if there is a battle between good and evil, that battle is a psychological one – it takes place in our heads, in the every-day, in the tiniest of decisions. Good is done when we fight against our negative behavioral predispositions and win. Bad's done when we're lazy, when we act to type. It's when we muster the strength to not behave like our fathers that we become better people, not when we confuse ourselves into begging some vengeful superhero that lives in the sky to deliver us from the will of some diabolical supernatural vice-monster. And as for religion's ideas of an afterlife of punishment or reward for good or bad behavior – we get our heaven and hell in this life, not in the next. We reincarnate ourselves – we change the content of our futures – every time we make a decision.

So after a year of searching, I've decided to try to pay more attention to each one of those daily decisions. Because I'm now convinced that there is evidence of something following death. Because ghosts exist. There really are such things as apparitions and EVP and poltergeists and heavy breathing in old rooms in the night. And humans, being human, feel compelled to explain that. But they can't. It's only the faithful who think they can. In this regard, Christians are just the same as witches and druids and anti-Satan vigilantes and sceptical monsterologists and hard rational scientists. They all think they've got answers, but really, they're all wildly theorising. The simple truth is – nobody knows. Nobody, not Dr Salter, Dr Garvey, Father Bill or the Founder, knows what happens when our brains finally flicker off. We're in the dark about death and the purpose of existence. And an awful lot of people, it seems, are scared of the dark. This is the thing that I've learned over the last twelve months about blind belief in the supernatural: faith is for the frightened. These are the things that scare humans more than anything else – death, loneliness and guilt. That's the ominous three, the holy trinity of dread. If you sign up for a supernatural belief like Christianity, these timeless problems disappear in a puff of incensed smoke. Death? No worries. Paradise awaits you. Lonely? Don't be daft – God loves you and is with you always. Guilt? Just say the word, and you will be forgiven.

And it's not just the Christians. There's a certain type of ghost-believer that's victim to this same syndrome. They use ghosts, just as Dr Salter said, to make themselves feel more important or to convince themselves their dead friends, family and lovers aren't just Spam for maggots. They use their cod logic to bring order and meaning to their chaotic and seemingly meaningless lives. And some of them use it to dress themselves up as instant experts. You can say anything you like about ghosts and, providing you do it with enough authority, you'll get your own slot on satellite TV.

But not all of the ghost-convinced are like this. Because if you strip away all the nonsense, you're left with something that most Christians will never have. You're left with evidence. Genuine, unexplained, skull-bucklingly fantastic evidence. For me, the extraordinary truth about ghosts doesn't lie in the individual experiences of one witness or another. It lies in the patterns. That, perhaps, four or five other people heard breathing in that bedroom before me doesn't make it four or five times more interesting, it makes it one of the most incredible mysteries in the world. Just like the previous occupants of Annie's room, the many victims of poltergeists, the worldwide thousands who've recorded EVP, the routinely spooked visitors to Michelham Priory, the young brothers who talked to the women in their bathroom, it's the chorus of humans who are experiencing the same things, evidence of intelligent ghosts, that make this subject so profound and wondrous and universal. I am convinced that one of the frontier sciences will eventually solve ghosts. And most likely it will be quantum theorists. With their atoms that don't like to be observed, their free-floating souls, their mysterious extra dimensions and their fundamental cosmic interconnectedness of all things, they certainly seem to be closing in on the subject. And, as for God, perhaps Lou's not so far from the truth when he describes him as a 'powerful energy source'. Perhaps this fundamental level that near-death survivors describe feeling so blissfully a part of is God – a Quantum God. Maybe it's this very essence of existence, this network of life, that we all have a pinch of fizzing away inside our microtubules, that's the sacred source that everybody's searching for. Or, then again, maybe not.

As for the hard sceptics, I think that to believe so passionately in the existence of nothing that isn't immediately obvious is to suffer the most gigantic failure of intelligence and imagination. The universe – the reality in which we exist – is such an immeasurable, unbelievable and, ultimately, unknowable thing. And the only thing *I* know for sure is that it's a stranger place than *any* human has the capacity to imagine.

For Denzel's mother, though, the universe is small and filled with Jesus. Here she is, one of the faithful frightened. Is she terrified that she might, in some way, be responsible for her son's disability? Or petrified that God won't judge her good enough to provide a cure? Our prayers and pep-talk over, Lou, Bob and I enter her house and find her nervous. There's a jittery hyperactivity crackling her blood and she clucks and laughs and works hard to fill the silences. In the night, her front room has taken on a sinister aspect. As Lou makes a quick tour of the house, ensuring that anything that's making a sound is identified and switched off at the mains, I look around. There's a Bible on the table, open at Ecclesiastes. I notice that all of the indoor shrubbery is made from plastic and there are little pottery praying angels, many of which have had their heads broken off. There's a torn scrap of paper pinned to the wall above the door. It says, in inky hand-writing, 'Whoever enters this house is covered by the blood of Jesus.'

Just as I'm reading this, Lou comes into the room. He's gone red.

'Guys,' he says, urgently. 'I need a word. Outside.'

Bob and I follow him out into the driveway. His face is glowing in the tungsten orange dark and the crickets are going crazy.

'Before,' he says, breathless from something, 'I told you that I took a pinch of blessed salt and I sprinkled it in front of the closet doorway, right? The one that the black shadow comes out of.'

'Yes,' I say.

'Well, I just walked in there and the room is in a fucking shambles. She's ripped the carpet up and bleached the fucking thing! You can smell the bleach!'

'She's bleached the area where you put the salt?' I say. This can't be right. The holy salt *cannot* have worked. 'Fuck! That's ... '

That's just like Father Amorth said. Bob looks aghast.

'That's like, if you don't like oregano,' says Bob, 'and you get the taste of oregano, you go ape.'

Ten minutes later, I'm sitting in Denzel's room, preparing for lights out. There are two EMF readers in front of the closet door. The kid's bed, chest of drawers and toy trunk have been pulled from the walls and left clumsily and uncomfortably out of place. The room is, indeed, a shambles. The carpet by the closet, where the blessed salt was sprinkled, has been pulled up and the air is bruised with the piercing, sweet smell of bleach. And there's the bottle, sitting nearby. Next to that, there's a wet and dirty cloth lying discarded on the floor, like a dead mouse. Pinned to the door of the wardrobe there's a scrap of paper that's been written on in green wax crayon. It says: 'The blood of Jesus covers this room.'

I sit still and wonder, quietly, until Lou's head comes round the door.

'You OK, Will?' he asks.

'Yeah,' I say.

'Remember,' he says. 'Rule number one. Don't freak out.'

'Yep,' I say. 'I remember.'

Lou switches the light off and closes the door. All I can see are the EMF meter's green LEDs glowing and a hazy halo of light pushing up through the gap above the blind. And then, surrounded by the thick and acrid darkness, I fall fast asleep.

Epilogue

Lou Gentile is working towards a pilot's licence, so that he can buy a light aircraft and help the demonically infestated all over America.

Kathy Ganiel is still using divination.

Trevor and Debbie from Avalon Skies got married, in secret, in Gretna Green.

The ever-sceptical Trevor requested a tape of Loping Coyote's 'Mary' possession in Tow Law, as he was having doubts about it. Loping Coyote resigned from the organisation soon after to 'pursue his Native American spirituality'. They apparently remain on speaking terms.

Charles and David have recently picked up signs of increased activity from Friends of Hecate in Clapham Woods. Their investigations continue.

Philip Hutchinson has a photograph of himself taken outside Michelham Priory, in which an unexplained mist is present.

A senior Ghost Club investigation squad, led by Lance, have been investigating Ham House in Richmond, Surrey. They have come up with some 'very interesting' results which will be published soon.

Stacey still lives with her hermit ghost under the stairs.

Lynne and Neil are getting used to living with Annie. She appeared, recently, looking worriedly out of a window. Her presence alerted Lynne to a fight that was taking place in the car park.

'Big' George eventually resigned from Ghosts-UK and joined an organisation called Club Zero. He soon became fed up with them too. He is currently a senior member of Jacquie's group Shanry.

Steve, the Trance Medium, claims to have been recently possessed by the ghost of King Henry VIII.

Following the publication of the first edition of this book, Charles Walker contacted Will to tell him that, during the Ouija board scene, the fact that he is a 'diabetic with bad circulation' wasn't taken into account.

Following the broadcast of the pilot episode of *Haunted Homes*, The Founder was droppped from the series.

Janet still lives quietly with her family in Clacton-on-Sea. Her youngest child still knows nothing of the events of 1977/78.

Father Amorth continues his gruelling schedule of exorcisms.

10 Most Haunted Places in America

Bachelor's Grove Cemetery, Chicago, Illinois

The now-derelict Bachelor's Grove Cemetery is notorious amongst paranormalists as being one of the most haunted corpse-parks in the world. Under the weeds and rubble of the ruined tombs lie the remains of Windy City residents dating back to 1844. Nobody has been buried here since 1965, when it was closed after falling into disrepair. The combined work of vandals, nature and local occultists have turned this small, one-acre location into the very definition of "spooky," with its cracked graves, gnarled bushes and bits of old candle, smashed crucifix and eviscerated virgin (probably) that local dabblers in the demonic have left behind. It's little wonder, then, that so much activity has been reported here. Most notably, a full female apparition who carries a baby in her arms (sometimes called the "Madonna of Bachelor's Grove"), a replay of a farmer being dragged by his horse and plow into the now-stagnant pond (which was, apparently, a favoured cadaver-dump for mobsters in the 1950s) and, weirdly, the ghost of a house which many people claimed to have seen whilst walking up the path that leads to the entrance.

Startling displays of ghost lights are also said to be common here, including red lights that dart away so fast they leave a trail and blue orbs that bounce from tombstone to tombstone.

Alcatraz Island, San Francisco Bay, California

Pity those poor Miwok Indians who were led, shackled and twitching with spasms of dread, onto Alcatraz Island in 1859 as the first residents of the prison. Not only had they been sentenced to serve time on what was to become one of the United States' most dismal penitentiaries, their particular tribe had feared the place for generations, convinced, as they were, that it was inhabited by evil spirits. And if the ghost chroniclers of San Francisco are to be believed, those wise old Native American elders might have been on to something. Alcatraz was turned from an army fort and prison into the largest reinforced concrete structure in the world in 1934. And, whether or not it was haunted in the days of the Miwok, many people claim that it is today, with the echoes of the inmates who were held there until its closure in 1963.

And they were a tormented people indeed. Alcatraz was the destination for America's most dangerous criminals, and they were sent to the lonely rock for the State to have its revenge. There was never even the pretence of rehabilitation. Prisoners were forbidden to talk, except for three minutes twice a day and two hours during the weekend as a special treat. Many, including Al Capone (who enjoyed playing his banjo, somewhat unaccountably, in the shower area), went mad; others were murdered or died from disease. Less ambitious types satisfied themselves with chopping off their own fingers with an axe. The guards were much more likely to beat you until you were a Picasso of body parts, bubbles of blood blowing out of each one of your five nostrils, than they were to deliver you a decent breakfast. The most feared area of the complex were the four solitary confinement "holes" in Block D – numbered 11-14. Inmates were kicked out, stripped and chucked into these concrete boxes with nothing but bread to eat and a hole to shit in, and the only thing they had to look forward to was a standard meal once every three days and, eventually, to being let out – back into the hellish warren of Alcatraz itself. Many, unsurprisingly, went comprehensively mental after a stretch in the hole. Rufe McCain didn't, though. He was forced to do an incredible three years and two months hole-time, after being caught trying to escape. And what did he do when he was eventually released? Keep his head down and his mouth shut (even during his three minutes chat-grace)? Make a grovelling apology to the chief warden? No, he found the man he was supposed to have escaped with – and he killed him.

Surprisingly, reports of supernatural oddery are not centred around Block D (with the exception, that is, of some ghost hunters feeling a little "strange" in the powerfully evocative little man-boxes – hardly unexplained, that). Rather, tour guides have reported hearing locks bolting, doors slamming, men shouting, screams and footsteps in corridors, all after the complex had closed for the day. Cell 14-D, where McCain was stored, is also, apparently, sometimes impossible to heat and the sound of banjo playing is heard in the shower area.

Hickory Hill, Equality, Illinois

The Old Slave House on Hickory Hill, near Harrisburg in southern Illinois, has had thoroughly grim history. It was built in 1842 by John Hart Crenshaw, a man who took ruthless advantage of a local law permitting the use of slaves to work in the salt mines of Saline County – an allowance that, at the time, was thought necessary, as nobody who wasn't in chains or acting under threat of torture and death would ever dream of taking a job down there, no matter what the pay and perks. But his wholehearted embracing of this nasty bit of legal footwork wasn't enough evil for the dastardly Crenshaw. He started kidnapping free African Americans and putting them to work down his salty holes and then selling spares to slave-owners in the South. And, when he ran out of excess humans, he started breeding them himself, using a stud known as 'Uncle Bob' (although you'd imagine 'Smiling Bob' might have been more apt). Bob is said to have fathered as many as 300 children and eventually passed away in 1948, at 112 years of age. The slaves were kept in the attic, which contained twelve cells and a whipping post. Each cell contained iron shackle-rings on the floor and tiny, barred windows. Ghostly activity often reported when the location was a tourist destination include spectral cries, whimpers and the sound of chinking chains. It's also reported that in the 1920s, an exorcist named Hickman Whittington visited the house and died some hours after leaving. In the 1960s, two Vietnam vets who tried to spend the night in the attic claimed they were surrounded by black shadows, and ran from the building, screaming. Soon after this, the owner stopped allowing visitors to stay after dark.

McLoughlin House, Oregon City, Oregon

Commonly known as the 'Father of Oregon,' John McLoughlin founded the city in 1829. By all accounts a wise and altruistic man, he gave away three hundred plots to needy settlers, schools and churches and was known to rescue pioneers who got themselves into trouble on the Oregon Trail. Despite all this, and his being a physician, mayor, councilman and a famously generous aid-giver, his wealth and Catholicism made him unpopular with the impoverished Protestant locals, and when Congress decided they disapproved of his claim to the land, he received little in the way of support from the ungrateful bastards. He died in 1857, a bitter and dejected man who felt betrayed by the world and, very possibly, with the concept of karma. With his impressive height of six foot five inches, his hollow eyes and his flowing white hair, McLoughlin already looked like a ghost, so it should have come as no surprise when, in the mid 1970s, strange things starting taking place in his old home. Sceptical curator Nancy Wilson felt a tap on her shoulder and, soon afterwards, several of her staff began reporting the sight of a tall and bulky black shadow walking along a corridor in one of the upper floors and disappearing into McLoughlin's old bedroom. In the same hall, footsteps have been heard and pipe tobacco has been smelled. Elsewhere in the old building, a child's bed sometimes appears slept in when staff do their rounds before opening in the morning, tassled lampshades are seen to move in unlikely motions, and sudden cries for help are heard as are loud, unexplained crashes.

The Lemp Mansion, St. Louis, Missouri

Formerly the home of nineteenth-century booze magnate William Lemp, Sr., and his mental family, this four-storey, thirty-four-room mansion is supposedly haunted to this day. William Sr.'s father, Johann Adam Lemp, arrived in Missouri from Germany in 1838 and set up business brewing vinegar and lager beer. He soon realised that the grateful locals found both of these products to be far tastier than the nasty English ale they'd been forced to drink up until then, but when they had the choice, their very favourite was the lager. So wise Johann stopped making vinegar altogether, and concentrated on the quick acquisition of a vast fizzy fortune. After his death in 1862, his son took over and, with impressive Donald Trump–style business cunning, soon turned the company into the world's largest brewery. At its peak, it was chucking out 900,000 barrels a year from a plant that covered eleven city blocks. All was spiffy with the Lemps, until a savage bad luck attack slashed across their collective lives, beginning in 1901 when, on a trip to Pasadena, the bright, elder son, Frederick, suffered a series of debilitating illnesses which ultimately led to his coronary arrest and death. His mortally saddened father reacted by sinking into a thick depression and, pushed on by the death of his best friend, Frederick Pabst, in 1904, shot himself through the heart with a small-calibre pistol.

The business then passed to flamboyant son William Jr., who reacted to the 1919 Prohibition laws by shrugging his shoulders, saying "Tch," closing the business and shooting himself with a small-calibre pistol. But not before his sister, Elsa, became depressed and shot herself with a small-calibre pistol. Then, in 1949, brother Charles walked into the mansion and shot his dog and then himself with a small cal – oh, you're waaaayy ahead of me.

The oppressive building was bought in the mid-seventies by Dick Pointer, Jr., who began turning it into a large restaurant and hotel. And this is when the haunting began in earnest. Horses' hooves were heard clattering up the long-grown-over path; workers' tools went missing, and some of the crew were so roundly spooked by the poltergeist activity that they ran out, never to return. When the renovation was complete, diners at the restaurant saw apparitions, and glasses hovering above the bar and flying through the air, and heard the sound of voices, doors locking and unlocking and a spectral piano.

Gettysburg, Pennsylvania

The battle that took place in and around the village of Gettysburg in July 1863 lasted for three days and was ruthless, bloody and hellish. Numerous ghosts are said to haunt the place nowadays, and there is space here to list but a few spectral highlights. On the official "battlefield" (the fighting, obviously, wasn't actually confined to this relatively small space) the sound of gunfire, shouting and weeping are often reported, as are sightings of fully uniformed apparitions. An entire battle was apparently witnessed by a group of visiting VIPs from Little Round Top. They assumed it was a bunch of local geeks having a reenactment, until they discovered that nobody was actually down there. The suitably named 'Devil's Den' was, like Alcatraz, thought to be infested with evil back in the days of the Native Americans. The invading Europeans didn't do much to help matters by killing a load of people there and leaving their corpses lying about the Den. Sightings are, perhaps unsurprisingly, notably frequent here. Other locations to check out include Rose Farm (phantom fights and ghosts by gravestones), Hummelbaugh House (howling dog), Pennsylvania Hall (groans, shouts, apparition of an entire makeshift operating theatre) and George Weikert House (tricky door).

The Brown Mountains, North Carolina

The Brown Mountains are famous for their ghost lights. Visible for miles around, the staggering displays of orbicular oddness have drawn the attention of parapsychologists and conventional scientists alike. According to respected American spectre expert Joshua P. Warren, they have been investigated three times by the U.S. government, once by the Weather Service and once by a team from the Smithsonian Institution, and all to no avail. The most convincing skeptical explanation – that they are the result of local geological faults scraping and sparking and perhaps creating something like ball lightning – seems less likely in light of reports of the orbs appearing in many different colours, dancing about, splitting into three or four smaller orbs which orbit around each other, or marching in straight lines along the ridge. Could these lights actually be the phantom remains of a violent battle that once took place in the area between Cherokee and Catawba Indians?

Winchester House, San Jose, California

Most houses become haunted through no fault of their own. Winchester House, however, was built by a lady (who I *wish* had been available for interview for this book ...) on the instructions of her dead husband (supposedly she chatted to him via Boston medium Adam Coons) for the sole purpose of being a sort of rooming-house for the dead. By the time she'd finished construction thirty-eight years later, Sarah Winchester had a house with seven hundred rooms, nine hundred and fifty doors and ten thousand windows over forty acres. The dark heart of this weird complex was The Blue Room, designed by dead spirits to be especially conducive to successful séances. Ms. Winchester would ring a bell in the tower three times a night to invite the floating souls in. Bad sorts were discouraged from entering by her cunning use of passageways with dead ends and various design paradiddles based around the number thirteen. Every so often, kindly Sarah would order her chefs and servants to lay on vast banquets, with solid gold plates and cutlery and five generous courses. Naturally, there were always thirteen places set and, also naturally, the only pre-mortal soul sat at the table would be Sarah herself. It is, perhaps, not a huge surprise to learn that Sarah was not a fan of actual living humans: she apparently sent both Theodore Roosevelt and Mary Baker Eddy on their way when they popped round for a tour. Harry Houdini was, however, permitted entry. He never spoke about the night he spent as a guest of Ms. Winchester and her invisible friends. Sarah passed away to join her mate in September 1922, and her dying wish, for "the ghosts to continue to be welcomed and provided for," was apparently honoured: ghost lights and a female apparition have been seen floating in the corridors, and whispers, slams and soft organ music is sometimes heard in the long-uninhabited rooms.

The Nicholson Mansion, Indianapolis, Indiana

Built in 1870 by a contractor for the Marion County Courthouse named David Nicholson, this impressive mansion was on the verge of demolition when, in 1997, it was saved by local enthusiasts and shifted, on two trailers, to a spare plot of land not far away. The town newspaper photographed the move and, when it published its story, was besieged by callers, all of whom had noticed what appeared to be the ghost of a girl gazing out of an upstairs window. Shortly after this, a policeman phoned up one of America's best known ghost hunters and asked him to have a poke about.

The house was still in two halves when Troy Taylor and his small team from the American Ghost Society arrived late in the afternoon. Two of them walked into the main section of the bisected building. They found nothing but a piano on the first floor, but on reaching a bedroom on the second floor, Taylor's EMF metre went berserk. This was especially strange as the house was completely unplugged – it had no electricity or water anywhere near it. Then, Taylor heard a shout. It was his colleague, Michael Barrett, calling from a back staircase. At the very moment Troy's EMF metre was hitting the red, a light that was hanging from the ceiling by a metal chain started swinging in a truly bizarre fashion. Taylor dashed to see what Barrett wanted and then the men watched in aghast fascination as the light swung back and forth, round in circles and even, every so often, stopped at an angle in midair for a few beats before swinging on again. Eventually, it came to an extremely sudden halt. To this day, those long minutes remain the most baffling and unexplainable of Taylor's notable fifteen-year career.

Pawleys Island, South Carolina

The story behind the sightings of a faceless grey man on Pawleys Island, off the coast of South Carolina, is as strange as it is unlikely. But, anyway, here goes.... Back in the eighteenth century, in the town of Charleston, there was a belle so beautiful that she could have had the pick of any of the bachelors in the neighbourhood. To the shame of her parents, however, she decided that out of all the men in Charleston, she wanted to marry her roguish cousin. Said cousin was then sent away to France, for fear of them shaming everyone with a litter of boss-eyed inbred babies. Despite his swearing solemnly that he would return and marry his true love, when he reached Europe, the cousin got himself involved in an unfortunate duel situation –

and, as everyone knows, if there are two things Frenchmen are good at, it is cheese and duels. The belle was bereft, but recovered soon enough to marry another local widower. They were happy, and got into the habit of spending malaria season on Pawleys Island, just off the coast, where the gentle breeze kept the mosquitos at bay.

One day in 1778, when her husband was off doing his part in the American Revolution, a hurricane sank a ship not far from the rocks of their summer island. That night, the sole survivor dragged himself onto dry land and to her front door, which she opened to discover ... the incest-hungry, heartbroken cousin. When he found out that she'd not waited for him, he fled, eventually dying of fever. The belle's husband returned, and they lived happily every after on their breezy bit of rock. Except, until the day that she died, she was troubled by regular sightings of a man who would lurk about the dunes and appear to be watching her.

Told you it was unlikely. But the phenomena that it seeks to explain is, apparently, all too real. The apparition of the man is still seen on the island whenever hurricanes are about to strike. Sightings have been reported before the storms of 1822, 1893, 1916, 1954 and 1955.

Sources

How to Hunt Ghosts, by Joshua P. Warren (Fireside); *The International Directory of Haunted Places*, by Dennis William Hauck (Penguin); *The Encyclopaedia of Ghosts and Spirits*, by Rosemary Ellen Guiley (Facts On File); *Confessions of a Ghost Hunter*, by Troy Taylor (Whitechapel Productions).

Acknowledgements

First and foremost I'd like to thank everybody who appears in this book, especially Maurice Grosse, Dr Mark Salter, Dr James Garvey, Lance Railton, Trevor and Debbie, Philip Hutchinson, David Vee, Charles Walker, Jacqueline Adair, Father Gabriele Amorth and, of course, the brilliant, brilliant Lou Gentile.

I'd also like to give a full royal bow to my editor Andrew Goodfellow and my agent Paul Moreton for always knowing what's best for me, even when I don't, and for putting up with my constant obsessive worrying.

Thanks also to Alex Hazle, Sarah Bennie and everyone at Ebury, Jon Ronson, Chloe Makin, Paul Merrill, Duncan Hayes, Francis Storr, Richard Purvis, Guy Lyon Playfair, Peter Johnson at the SPR, Andrew Sumner, Simon Hills, Jill Schwartzman and all at Harper paperback in New York, Antony Medley, Simon Trewin and Danny Wallace.